Chemical Engineering

Chemical Engineering

Rob Hanson

Larsen & Keller
www.larsen-keller.com

Chemical Engineering
Rob Hanson
ISBN: 978-1-64172-428-9 (Hardback)

Larsen & Keller
Published by Larsen and Keller Education,
5 Penn Plaza,
19th Floor,
New York, NY 10001, USA

Cataloging-in-Publication Data

Chemical engineering / Rob Hanson.
p. cm.
Includes bibliographical references and index.
ISBN 978-1-64172-428-9
1. Chemical engineering. 2. Engineering. 3. Chemistry, Technical.
4. Metallurgy. I. Hanson, Rob.
TP155 .C44 2020
660--dc23

Printed and bound in China.

For more information regarding Larsen and Keller Education and its products, please visit the publisher's website www.larsen-keller.com

TABLE OF CONTENTS

PREFACE

It is with great pleasure that I present this book. It has been carefully written after numerous discussions with my peers and other practitioners of the field. I would like to take this opportunity to thank my family and friends who have been extremely supporting at every step in my life.

The branch of engineering that uses the principles of chemistry, physics, biology, mathematics and economics for designing, production, transportation and transformation of energy and materials is referred to as chemical engineering. It is involved in plant design and operation including the assessment of safety and hazards, process design and analysis. It is also closely associated with control engineering, chemical reaction engineering, biological engineering, and nuclear engineering. Chemical engineering develops economic ways of using materials and energy. It includes chemistry and engineering that convert raw materials into usable products such as medicine, petrochemicals, and plastics. This textbook traces the progress of this field and explores some of its key concepts and applications. While understanding the long-term perspectives of the topics, it makes an effort in highlighting their impact as a modern tool for the growth of the discipline. This book, with its detailed analyses and data, will prove immensely beneficial to professionals and students involved in this area at various levels.

The chapters below are organized to facilitate a comprehensive understanding of the subject:

Chapter - What is Chemical Engineering?

The branch of chemistry that is concerned with the efficient use, production, design and transformation of energy and materials is known as chemical engineering. It uses the principles of physics, chemistry, mathematics, biology and economics. The topics elaborated in this chapter will help in gaining a better perspective about chemical engineering.

Chapter - Branches of Chemical Engineering

There are various sub-disciplines of chemical engineering. These can be categorized into polymer engineering, chemical reaction engineering, biochemical engineering, pharmaceutical engineering, etc. This chapter closely examines these sub-disciplines of chemical engineering to provide an extensive understanding of the subject.

Chapter - Chemical Kinetics

Chemical kinetics is defined as the branch of chemistry which deals with the under-standing of the rates of chemical reactions. A few of its significant areas are collision theory, chemical equilibrium, law of mass action, etc. This chapter discusses in detail all these aspects related to chemical kinetics.

Chapter - Electrochemical Engineering

The branch of chemical engineering that deals with the applications of various electrochemical phenomena such as electrowinning and refining, electrochemical separations and corrosion is known as electrochemical engineering. This chapter has been carefully written to provide an easy understanding about electrochemical engineering and electrochemical cells.

Chapter - Diverse Aspects of Chemical Engineering

Some of the fundamental aspects of chemical engineering include separation process¬es, unit operations, chemical kinetics, process design, etc. The topics elaborated in this chapter will help in gaining a better perspective about these processes of chemical engineering.

Rob Hanson

What is Chemical Engineering?

1

- **Chemical Element**
- **Chemical Compound**

The branch of chemistry that is concerned with the efficient use, production, design and transformation of energy and materials is known as chemical engineering. It uses the principles of physics, chemistry, mathematics, biology and economics. The topics elaborated in this chapter will help in gaining a better perspective about chemical engineering.

Chemical engineering is the development of processes and the design and operation of plants in which materials undergo changes in their physical or chemical state. Applied throughout the process industries, it is founded on the principles of chemistry, physics, and mathematics.

The laws of physical chemistry and physics govern the practicability and efficiency of chemical engineering operations. Energy changes, deriving from thermodynamic considerations, are particularly important. Mathematics is a basic tool in optimization and modeling. Optimization means arranging materials, facilities, and energy to yield as productive and economical an operation as possible. Modeling is the construction of theoretical mathematical prototypes of complex process systems, commonly with the aid of computers.

Chemical engineering is as old as the process industries. Its heritage dates from the fermentation and evaporation processes operated by early civilizations. Modern chemical engineering emerged with the development of large-scale, chemical-manufacturing operations in the second half of the 19th century. Throughout its development as an independent discipline, chemical engineering has been directed toward solving problems of designing and operating large plants for continuous production.

Manufacture of chemicals in the mid-19th century consisted of modest craft operations. Increase in demand, public concern at the emission of noxious effluents, and competition between rival processes provided the incentives for greater efficiency. This led to the emergence of combines with resources for larger operations and caused the transition from a craft to a science-based industry. The result was a demand for chemists with knowledge of manufacturing processes, known as industrial chemists or chemical technologists. The term chemical engineer was in general use by about 1900. Despite its emergence in traditional chemicals manufacturing, it was through its role in the development of the petroleum industry that chemical engineering became firmly established as a unique discipline. The demand for plants capable of operating physical separation processes continuously at high levels of efficiency was a challenge that could not be met by the traditional chemist or mechanical engineer.

Chemical

A chemical is any substance consisting of matter. This includes any liquid, solid, or gas. A chemical is any pure substance (an element) or any mixture (a solution, compound, or gas).

Examples of Naturally-occurring Chemicals

Naturally-occurring chemicals can be solid, liquid, or gas. Naturally occurring solids, liquids, or gases may be made up of individual elements or may contain many elements in the form of molecules.

- Gases: Oxygen and nitrogen are naturally-occurring gases. Together, they make up most of the air we breathe. Hydrogen is the most common naturally-occurring gas in the universe.
- Liquids: Perhaps the most important naturally-occurring liquid in the universe is water. Made up of hydrogen and oxygen, water behaves differently from most other liquids because it expands when frozen. This natural chemical behavior has had a profound effect on the geology, geography, and biology of Earth and (almost certainly) other planets.
- Solids: Any solid object found in the natural world is made up of chemicals. Plant fibers, animal bones, rocks, and soil are all made up of chemicals. Some minerals, such as copper or zinc, are made entirely from one element. But granite, for example, is a metamorphic rock made up of multiple elements.

Examples of Artificially Made Chemicals

Human beings probably began combining chemicals before recorded history. About 5,000 years ago, we know that people began combining metals (copper and tin) to create a strong, malleable metal called bronze. The invention of bronze was a major event, as it made it possible to form a huge range of new tools, weapons, and armor.

Bronze is an alloy (a combination of multiple metals and other elements), and alloys have become a staple of construction and trade. Over the past few hundred years, many different combinations of elements have resulted in the creation of stainless steel, lightweight aluminum, foils, and other very useful products.

Artificial chemical compounds have transformed both the food and pharmaceutical industry. Combinations of elements have made it possible to preserve and flavor food inexpensively, and chemicals are also used to create a range of textures from crunchy to chewy to smooth. Artificial chemical compounds are a major part of the pharmaceutical industry. By combining active and inactive chemicals in pills, pharmacists can treat many different disorders.

Chemical Element

Chemical element, also called element is any substance that cannot be decomposed into simpler substances by ordinary chemical processes. Elements are the fundamental materials of which all matter is composed.

At present there are 118 known chemical elements. About 20 percent of them do not exist in nature (or are present only in trace amounts) and are known only because they have been synthetically prepared in the laboratory. Of the known elements, 11 (hydrogen, nitrogen, oxygen, fluorine, chlorine, and the six noble gases) are gases under ordinary conditions, two (bromine and mercury) are liquids (two more, cesium and gallium, melt at about or just above room temperature), and the rest are solids. Elements can combine with one another to form a wide variety of more complex substances called compounds. The number of possible compounds is almost infinite; perhaps a million are known, and more are being discovered every day. When two or more elements combine to form a compound, they lose their separate identities, and the product has characteristics quite different from those of the constituent elements. The gaseous elements hydrogen and oxygen, for example, with quite different properties, can combine to form the compound water, which has altogether different properties from either oxygen or hydrogen. Water clearly is not an element because it consists of, and actually can be decomposed chemically into, the two substances hydrogen and oxygen; these two substances, however, are elements because they cannot be decomposed into simpler substances by any known chemical process. Most samples of naturally occurring matter are physical mixtures of compounds. Seawater, for example, is a mixture of water and a large number of other compounds, the most common of which is sodium chloride, or table salt. Mixtures differ from compounds in that they can be separated into their component parts by physical processes; for example, the simple process of evaporation separates water from the other compounds in seawater.

Atomic Nature of the Elements

Paralleling the development of the concept of elements was an understanding of the nature of matter. At various times in history, matter has been considered to be either continuous or discontinuous. Continuous matter is postulated to be homogeneous and divisible without limit, each part exhibiting identical properties regardless of size. This was essentially the point of view taken by Aristotle when he associated his elemental qualities with continuous matter. Discontinuous matter, on the other hand, is conceived of as particulate—that is, divisible only up to a point, the point at which certain basic units called atoms are reached. According to this concept, also known as the atomic hypothesis, subdivision of the basic unit (atom) could give rise only to particles with profoundly different properties. Atoms, then, would be the ultimate carriers of the properties associated with bulk matter.

The atomic hypothesis is usually credited to the Greek philosopher Democritus, who considered all matter to be composed of atoms of the four elements—earth, air, fire, and water. But Aristotle's concept of continuous matter generally prevailed and influenced thought until experimental findings in the 16th century forced a return to the atomic theory. Two types of experimental evidence gave support to the atomic hypothesis: first, the detailed behaviour of gaseous substances and, second, the quantitative weight relationships observed with a variety of chemical reactions. The English chemist John Dalton was the first to explain the empirically derived laws of chemical combination by postulating the existence of atoms with unique sets of properties. At the time, chemical combining power (valence) and relative atomic weights were the properties of most interest. Subsequently numerous independent experimental verifications of the atomic hypothesis were carried out, and today it is universally accepted. Indeed, in 1969 individual uranium and thorium atoms were actually observed by means of an electron microscope.

The Structure of Atoms

Atoms of elemental substances are themselves complex structures composed of more fundamental particles called protons, neutrons, and electrons. Experimental evidence indicates that, within an atom, a small nucleus, which generally contains both protons and neutrons, is surrounded by a swarm, or cloud, of electrons. The fundamental properties of these subatomic particles are their weight and electrical charge. Whereas protons carry a positive charge and electrons a negative one, neutrons are electrically neutral. The diameter of an atom (about 10^{-8} centimetre) is 10,000 times larger than that of its nucleus. Neutrons and protons, which are collectively called nucleons, have relative weights of approximately one atomic mass unit, whereas an electron is only about $^{1}/_{2000}$ as heavy. Because neutrons and protons occur in the nucleus, virtually all of the mass of the atom is concentrated there. The number of protons in the nucleus is equivalent to the atomic number of the element. The total number of protons and neutrons is called the mass number because it equals the relative weight of that atom compared to other atoms. Because the atom itself is electrically neutral, the atomic number represents not only the number of protons, or positive charges, in the nucleus but also the number of electrons, or negative charges, in the extranuclear region of the atom.

The chemical characteristics of elements are intimately related to the number and arrangement of electrons in their atoms. Thus, elements are completely distinguishable from each other by their atomic numbers. The realization that such is the case leads to another definition of an element, namely, a substance, all atoms of which have the same atomic number.

The Existence of Isotopes

Careful experimental examination of naturally occurring samples of many pure elements shows that not all the atoms present have the same atomic weight, even though they all have the same atomic number. Such a situation can occur only if the atoms have different numbers of neutrons in their nuclei. Such groups of atoms—with the same atomic number but with different relative weights—are called isotopes. The number of isotopic forms that a naturally occurring element possesses ranges from one (e.g., fluorine) to as many as ten (e.g., tin); most of the elements have at least two isotopes. The atomic weight of an element is usually determined from large numbers of atoms containing the natural distribution of isotopes, and, therefore, it represents the average isotopic weight of the atoms constituting the sample. More recently, precision mass-spectrometric methods have been used to determine the distribution and weights of isotopes in various naturally occurring samples of elements.

Chemical Compound

Chemical compound is any substance composed of identical molecules consisting of atoms of two or more chemical elements.

All the matter in the universe is composed of the atoms of more than 100 different chemical elements, which are found both in pure form and combined in chemical compounds. A sample of any given pure element is composed only of the atoms characteristic of that element, and the atoms of each element are unique. For example, the atoms that constitute carbon are different from those that make up iron, which are in turn different from those of gold. Every element is designated by a unique symbol consisting of one, two, or three letters arising from either the current element name or its original name. For example, the symbols for carbon, hydrogen, and oxygen are simply C, H, and O, respectively. The symbol for iron is Fe, from its original Latin name ferrum. The fundamental principle of the science of chemistry is that the atoms of different elements can combine with one another to form chemical compounds. Methane, for example, which is formed from the elements carbon and hydrogen in the ratio four hydrogen atoms for each carbon atom, is known to contain distinct CH_4 molecules. The formula of a compound—such as CH_4—indicates the types of atoms present, with subscripts representing the relative numbers of atoms (although the numeral 1 is never written).

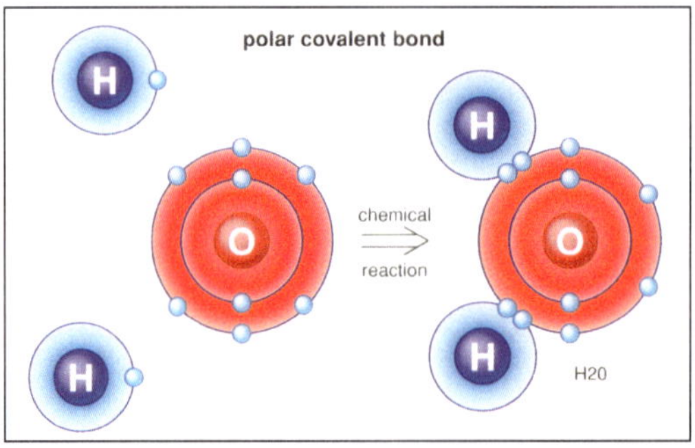

A water molecule is made up of two hydrogen atoms and one oxygen atom. A single oxygen atom contains six electrons in its outer shell, which can hold a total of eight electrons. When two hydrogen atoms are bound to an oxygen atom, the outer electron shell of oxygen is filled.

Water, which is a chemical compound of hydrogen and oxygen in the ratio two hydrogen atoms for every oxygen atom, contains H_2O molecules. Sodium chloride is a chemical compound formed from sodium (Na) and chlorine (Cl) in a 1:1 ratio. Although the formula for sodium chloride is NaCl, the compound does not contain actual NaCl molecules. Rather, it contains equal numbers of sodium ions with a charge of positive one (Na^+) and chloride ions with a charge of negative one (Cl^-). The substances mentioned above exemplify the two basic types of chemical compounds: molecular (covalent) and ionic. Methane and water are composed of molecules; that is, they are molecular compounds. Sodium chloride, on the other hand, contains ions; it is an ionic compound.

The atoms of the various chemical elements can be likened to the letters of the alphabet: just as the letters of the alphabet are combined to form thousands of words, the atoms of the elements can combine in various ways to form a myriad of compounds. In fact, there are millions of chemical compounds known, and many more millions are possible but have not yet been discovered or synthesized. Most substances found in nature—such as wood, soil, and rocks—are mixtures of chemical compounds. These substances can be separated into their constituent compounds by physical methods, which are methods that do not change the way in which atoms are aggregated within the compounds. Compounds can be broken down into their constituent elements by chemical changes. A chemical change (that is, a chemical reaction) is one in which the organization of the atoms is altered. An example of a chemical reaction is the burning of methane in the presence of molecular oxygen (O_2) to form carbon dioxide (CO_2) and water.

$$CH_4 + 2O_2 \rightarrow CO_2 + 2H_2O$$

In this reaction, which is an example of a combustion reaction, changes occur in the way that the carbon, hydrogen, and oxygen atoms are bound together in the compounds.

Chemical compounds show a bewildering array of characteristics. At ordinary temperatures and pressures, some are solids, some are liquids, and some are gases. The colours of the various compounds span those of the rainbow. Some compounds are highly toxic to humans, whereas others are essential for life. Substitution of only a single atom within a compound may be responsible for changing the colour, odour, or toxicity of a substance. So that some sense can be made out of this great diversity, classification systems have been developed. An example cited above classifies compounds as molecular or ionic. Compounds are also classified as organic or inorganic. Organic compounds, so called because many of them were originally isolated from living organisms, typically contain chains or rings of carbon atoms. Because of the great variety of ways that carbon can bond with itself and other elements, there are more than nine million organic compounds. The compounds that are not considered to be organic are called inorganic compounds.

Mercury (Chemical Symbol: Hg) is the only metal element that is liquid at room temperature.

Within the broad classifications of organic and inorganic are many subclasses, mainly based on the specific elements or groups of elements that are present. For example, among the inorganic compounds, oxides contain O^{2-} ions or oxygen atoms, hydrides contain H^- ions or hydrogen atoms, sulfides contain S^{2-} ions, and so forth. Subclasses of organic compounds include alcohols (which contain the $-OH$ group), carboxylic acids (characterized by the $-COOH$ group), amines (which have an $-NH_2$ group), and so on.

The Periodic Table

The different abilities of the various atoms to combine to form compounds can best be understood in terms of the periodic table. The periodic table was originally constructed to represent the patterns observed in the chemical properties of the elements. That is to say, as the science of chemistry developed, it was observed that elements could be grouped according to their chemical reactivity. Elements with similar properties are listed in vertical columns of the periodic table and are called groups. As the details of the atomic structure were revealed, it became clear that the position of an element in the periodic table correlates with the arrangement of the electrons possessed by the atoms of that element. In particular, it was observed that the electrons that determine the chemical behaviour of an atom are those in its outermost shell. Such electrons are called valence electrons.

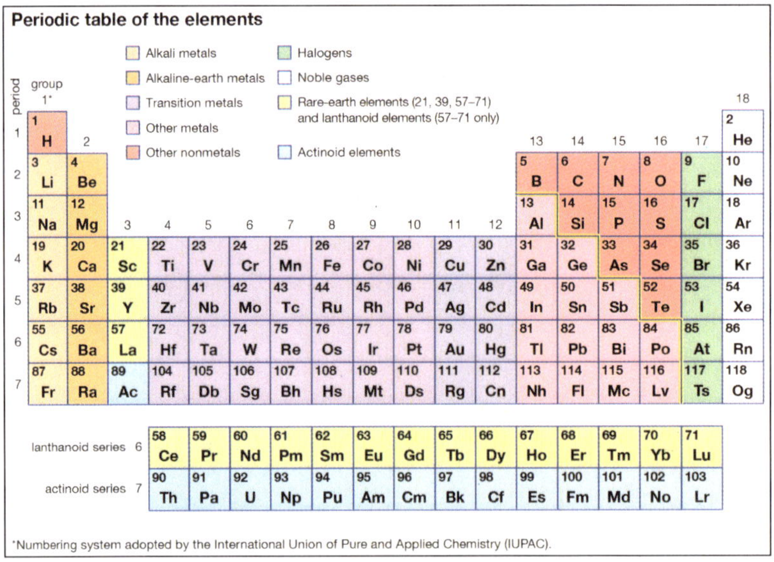

The periodic table of the elements.

For example, the atoms of the elements in Group 1 of the periodic table all have one valence electron, the atoms of the elements in Group 2 have two valence electrons, and so on, until Group 18, whose elements contain eight valence electrons, is reached. The simplest and most important rule for predicting how atoms form compounds is that atoms tend to combine in ways that allow them either to empty their valence shell or to complete it (i.e., fill it up), in most cases with a total of eight electrons. Elements on the left side of the periodic table tend to lose their valence electrons in chemical reactions. Sodium (in Group 1), for example, tends to lose its lone valence electron to form an ion with a charge of +1. Each sodium atom has 11 electrons (e^-), each with a charge of −1, to just balance the +11 charge on its nucleus. Losing one electron leaves it with 10 negative charges and 11 positive charges to give a net +1 charge: $Na \rightarrow Na^+ + e^-$. Potassium, located directly beneath sodium in Group 1, also forms +1 ions (K^+) in its reactions, as do the remaining members of Group 1: rubidium (Rb), cesium (Cs), and francium (Fr).

The atoms of the elements toward the right end of the periodic table tend to undergo reactions such that they gain (or share) enough electrons to complete their valence shell. For example, oxygen in Group 16 has six valence electrons and thus needs two more electrons to complete its outermost shell. Oxygen achieves this arrangement by reacting with elements that can lose or share electrons. An oxygen atom, for instance, can react with a magnesium (Mg) atom (in Group 2) by taking magnesium's two valence electrons, producing Mg^{2+} and O^{2-} ions. (When a neutral magnesium atom loses two electrons, it forms the Mg^{2+} ion, and, when a neutral oxygen atom gains two electrons, it forms the O^{2-} ion.) The resulting Mg^{2+} and O^{2-} then combine in a 1:1 ratio to give the ionic compound MgO (magnesium oxide). (Although the compound magnesium oxide contains charged species, it has no net charge, because it contains equal numbers of Mg^{2+} and O^{2-} ions.) Likewise, oxygen reacts with calcium (just below magnesium in Group 2) to form CaO (calcium oxide). Oxygen reacts in a similar way with beryllium

(Be), strontium (Sr), barium (Ba), and radium (Ra), the remaining elements in Group 2. The key point is that, because all the elements in a given group have the same number of valence electrons, they form similar compounds.

The chemical elements can be classified in many different ways. The most fundamental division of the elements is into metals, which constitute the majority of the elements, and nonmetals. The typical physical properties of metals are lustrous appearance, malleability (the ability to be pounded into a thin sheet), ductility (the ability to be drawn into a wire), and efficient thermal and electrical conductivity. The most important chemical property of metals is the tendency to give up electrons to form positive ions. Copper (Cu), for example, is a typical metal. It is lustrous but easily tarnishes; it is an excellent conductor of electricity and is commonly used for electrical wires; and it is readily formed into products of various shapes, such as pipes for water systems. Copper is found in many ionic compounds in the form of either the Cu^{+} or the Cu^{2+} ion.

The metallic elements are found on the left side and in the centre of the periodic table. The metals of Groups 1 and 2 are called the representative metals; those in the centre of the periodic table are called the transition metals. The lanthanoids and actinoids shown below the periodic table are special classes of transition metals.

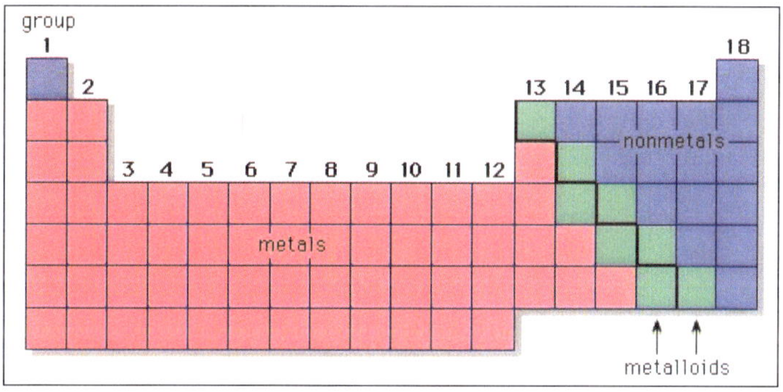

Metals, nonmetals, and metalloids are represented in different regions of the periodic table.

The nonmetals, which are relatively few in number, are found in the upper right-hand corner of the periodic table—except for hydrogen, the only nonmetallic member of Group 1. The physical properties characteristic of metals are absent in nonmetals. In chemical reactions with metals, nonmetals gain electrons to form negative ions. Nonmetallic elements also react with other nonmetals, in this case forming molecular compounds. Chlorine is a typical nonmetal. At ordinary temperatures, elemental chlorine contains Cl_2 molecules and reacts with other nonmetals to form molecules such as HCl, CCl_4, and PCl_3. Chlorine reacts with metals to form ionic compounds containing Cl^{-} ions.

The division of the elements into metals and nonmetals is only approximate. A few elements along the dividing line exhibit both metallic and nonmetallic properties and are called metalloids, or semimetals.

Trends in the Chemical Properties of the Elements

Reactivity series of metals	
Potassium	Most reactive
Sodium	
Calcium	
Magnesium	
Aluminum	
Zinc	
Iron	
Tin	
Lead	
Copper	
Silver	
Gold	
Platinum	Least reactive

As mentioned above, the characteristic chemical property of a metal atom is to lose one or more of its electrons to form a positive ion. However, certain metals lose electrons much more readily than others. In particular, cesium (Cs) can give up its valence electron more easily than can lithium (Li). In fact, for the alkali metals (the elements in Group 1), the ease of giving up an electron varies as follows:

$$Cs > Rb > K > Na > Li$$

with Cs the most likely, and Li the least likely, to lose an electron. In going down the group, the metals become more likely to lose an electron because the electron being removed lies increasingly farther from the positive nucleus. That is, the electron lost from Cs to form Cs^+ lies at a much greater distance from the attractive positive nucleus—and is thus easier to remove—than the electron that must be removed from a lithium atom to form Li^+. The same trend also is seen among the Group 2 elements (the alkaline-earth metals); the farther down in the group the metal resides, the more likely it is to lose an electron.

Just as metals vary somewhat in their properties, so do nonmetals. As a general rule, the most chemically active metals appear in the lower left-hand region of the periodic table, whereas the most chemically active nonmetals appear in the upper right-hand region. The properties of the semimetals, or metalloids, lie between those of the metals and the nonmetals.

The ionization energy of an element is the energy required to remove an electron from an individual atom (here M(g) represents a metal in the vapour state).

$$M(g) \xrightarrow{\text{ionization energy}} M^+(g) + e^-$$

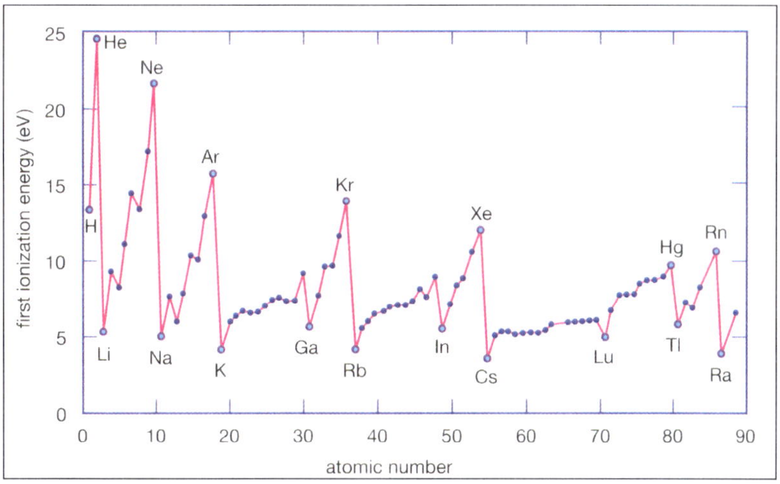

The ionization energy of an element is the energy required for removing an electron from an individual atom. Metal atoms tend to have low ionization energies relative to nonmetal atoms.

Metal atoms lose electrons to nonmetal atoms because metals typically have relatively low ionization energies. Metals at the bottom of a group lose electrons more easily than those at the top. That is, ionization energies tend to decrease in going from the top to the bottom of a group. Nonmetals, which are found in the right-hand region of the periodic table, have relatively large ionization energies and therefore tend to gain electrons. Ionization energies generally increase in going from left to right across a given period. Thus, the elements that appear in the lower left-hand region of the periodic table have the lowest ionization energies (and are therefore the most chemically active metals), while the elements that occur in the upper right-hand region of the periodic table have the highest ionization energies (and are thus the most chemically active nonmetals).

As mentioned above, when a nonmetallic element reacts with a metallic element, electrons are transferred from the atoms of the metal to the atoms of the nonmetal, forming positive ions (cations) and negative ions (anions), respectively. This produces an ionic compound. For example, lithium and fluorine (F) react to form lithium fluoride (LiF), which contains Li^+ and F^- ions.

In contrast, when two nonmetallic elements react, the atoms combine to form molecules by sharing electrons. Bonds formed by electron sharing between atoms are called covalent bonds. The electrons are shared rather than transferred, because the two nonmetal atoms have comparable attractive powers for the electrons in the bond. For example, fluorine gas consists of F_2 molecules in which the fluorine atoms are bound together by sharing a pair of electrons, one contributed by each atom. In addition, hydrogen and fluorine react to form hydrogen fluoride, which contains HF molecules. The hydrogen and fluorine atoms are bound together by a pair of electrons, one electron contributed by the hydrogen atom and one by the fluorine atom. Although the electrons are shared between the hydrogen and the fluorine atoms, in this case they are not shared equally. This is clear from the fact that the HF molecule is polar; the hydrogen atom has a partial positive charge (δ+), while the fluorine atom has a partial negative charge (δ−):

H−F

δ + δ−

(In this example the symbol δ stands for a number less than one.) This electrical polarity occurs because the shared electrons spend more time close to the fluorine atom than to the hydrogen atom. That is, fluorine has greater affinity for the shared electrons than does hydrogen. This leads to a polar covalent bond.

The ability of an atom to attract the electrons shared with another atom is termed its electronegativity. The relative electronegativities of the various atoms can be determined by measuring the polarities of the bonds involving the atoms in question. Fluorine has the greatest electronegativity value (4.0, according to the Pauling scale), and cesium and francium have the smallest values (0.79 and 0.7, respectively). In general, nonmetal atoms have higher electronegativities than metal atoms. In the periodic table, electronegativity typically increases in moving across a period and decreases in going down a group. When elements with very different electronegativities (such as fluorine and cesium) react, one or more electrons are transferred to form an ionic compound. For example, cesium and fluorine react to form CsF, which contains Cs^+ and F^- ions. When nonmetal atoms with differing electronegativities react, they form molecules with polar covalent bonds.

increasing electronegativity →

decreasing electronegativity ↓

group 1	2	3	4	5	6	7	8	9	10	11	12	13	14	15	16	17	18
H 2.2																	He –
Li 1.0	Be 1.6											B 2.0	C 2.6	N 3.0	O 3.4	F 4.0	Ne –
Na 0.9	Mg 1.3											Al 1.6	Si 1.9	P 2.2	S 2.6	Cl 3.2	Ar –
K 0.8	Ca 1.0	Sc 1.4	Ti 1.5	V 1.6	Cr 1.7	Mn 1.6	Fe 1.8	Co 1.9	Ni 1.9	Cu 1.9	Zn 1.7	Ga 1.8	Ge 2.0	As 2.2	Se 2.6	Br 3.0	Kr –
Rb 0.8	Sr 1.0	Y 1.2	Zr 1.3	Nb 1.6	Mo 2.2	Tc 2.1	Ru 2.2	Rh 2.3	Pd 2.2	Ag 1.9	Cd 1.7	In 1.8	Sn 2.0	Sb 2.1	Te 2.1	I 2.7	Xe 2.6
Cs 0.8	Ba 0.9	La-Lu 1.1-1.3	Hf 1.3	Ta 1.5	W 1.7	Re 1.9	Os 2.2	Ir 2.2	Pt 2.2	Au 2.4	Hg 1.9	Tl 1.8	Pb 1.8	Bi 1.9	Po 2.0	At 2.2	Rn –
Fr 0.7	Ra 0.9	Ac-Lr 1.1-1.7	104 –	105 –	106 –	107 –	108 –	109 –	110 –	111 –	112 –						

Each element has an electronegativity value, which is a measure of the ability of an atom to attract and share electron pairs of another atom.

Another important atomic property is atomic size. The sizes of atoms vary; atoms generally tend to become larger in going down a group on the periodic table and smaller in going from left to right across a period.

Classification of Compounds

Chemical compounds may be classified according to several different criteria. One common method is based on the specific elements present. For example, oxides contain one or more oxygen atoms, hydrides contain one or more hydrogen atoms, and halides contain one or more halogen (Group 17) atoms. Organic compounds are characterized as those compounds with a backbone of carbon atoms, and all the remaining compounds are classified as inorganic. As the name suggests, organometallic compounds are organic compounds bonded to metal atoms.

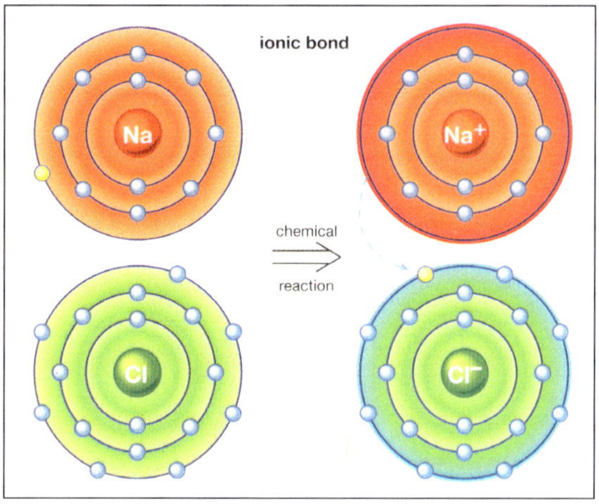

Ionic bond: sodium chloride, or table salt.

Ionic bonding in sodium chloride. An atom of sodium (Na) donates one of its electrons to an atom of chlorine (Cl) in a chemical reaction, and the resulting positive ion (Na^+) and negative ion (Cl^-) form a stable ionic compound (sodium chloride; common table salt) based on this ionic bond.

Another classification scheme for chemical compounds is based on the types of bonds that the compound contains. Ionic compounds contain ions and are held together by the attractive forces among the oppositely charged ions. Common salt (sodium chloride) is one of the best-known ionic compounds. Molecular compounds contain discrete molecules, which are held together by sharing electrons (covalent bonding). Examples are water, which contains H_2O molecules; methane, which contains CH_4 molecules; and hydrogen fluoride, which contains HF molecules.

A third classification scheme is based on reactivity—specifically, the types of chemical reactions that the compounds are likely to undergo. For example, acids are compounds that produce H^+ ions (protons) when dissolved in water to produce aqueous solutions. Thus, acids are defined as proton donors. The most common acids are aqueous solutions of HCl (hydrochloric acid), H_2SO_4 (sulfuric acid), HNO_3 (nitric acid), and H_3PO_4 (phosphoric acid). Bases, on the other hand, are proton acceptors. The most common base is the hydroxide ion (OH^-), which reacts with an H^+ ion to form a water molecule.

$$H^+ + OH^- \rightarrow HOH \text{ (usually written } H_2O\text{)}$$

Oxidation-reduction reactions constitute another important class of chemical reactions. Oxidation involves a loss of electrons, whereas reduction involves a gain of electrons.

For example, in the reaction between sodium metal and chlorine gas to form sodium chloride,

$$2Na + Cl_2 \rightarrow 2NaCl,$$

Electrons (e^-) are transferred from sodium atoms to chlorine atoms to form Na^+ and Cl^- ions in the reaction product, sodium chloride.

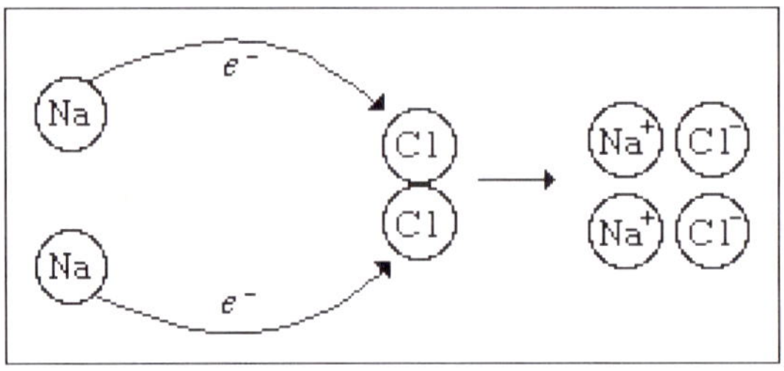

In this process, each sodium atom loses an electron and is thus oxidized, and each chlorine atom gains an electron and is thus reduced. In this reaction, sodium is called the reducing agent (it furnishes electrons), and chlorine is called the oxidizing agent (it consumes electrons). The most common reducing agents are metals, for they tend to lose electrons in their reactions with nonmetals. The most common oxidizing agents are halogens—such as fluorine (F_2), chlorine (Cl_2), and bromine (Br_2)—and certain oxy anions, such as the permanganate ion (MnO_4^-) and the dichromate ion ($Cr_2O_7^{2-}$).

Inorganic Compounds

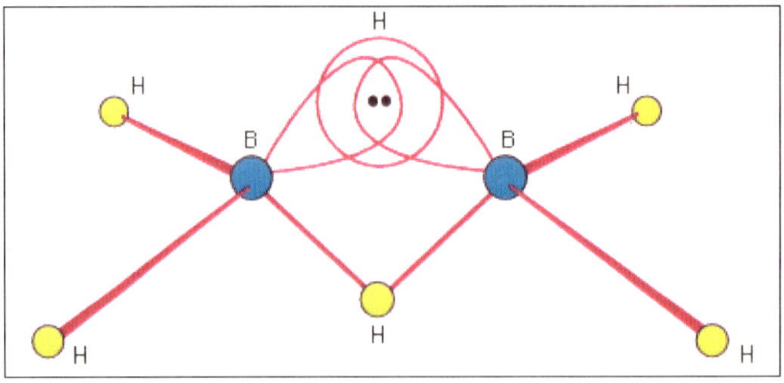

Borane is an example of an inorganic compound.

Inorganic compounds include compounds that are made up of two or more elements other than carbon, as well as certain carbon-containing compounds that lack carbon-carbon bonds, such as cyanides and carbonates. Inorganic compounds are most often classified in terms of the elements or groups of elements that they contain. Oxides, for example, can be either ionic or molecular. Ionic oxides contain O^{2-} (oxide) ions and metal cations, whereas molecular oxides contain molecules in which oxygen (O) is covalently bonded to other nonmetals such as sulfur (S) or nitrogen (N). When ionic oxides are dissolved in water, the O^{2-} ions react with water molecules to form hydroxide ions (OH^-), and a basic solution results. Molecular oxides react with water to produce oxyacids, such as sulfuric acid (H_2SO_4) and nitric acid (HNO_3). In addition, inorganic compounds include hydrides (containing hydrogen atoms or H^- ions), nitrides (containing N^{3-} ions), phosphides (containing P^{3-} ions), and sulfides (containing S^{2-} ions).

Transition metals form a great variety of inorganic compounds. The most important of these are coordination compounds in which the metal atom or ion is surrounded by two to six ligands. Ligands are ions or neutral molecules with electron pairs that they can donate to the metal atom to form a coordinate-covalent bond.

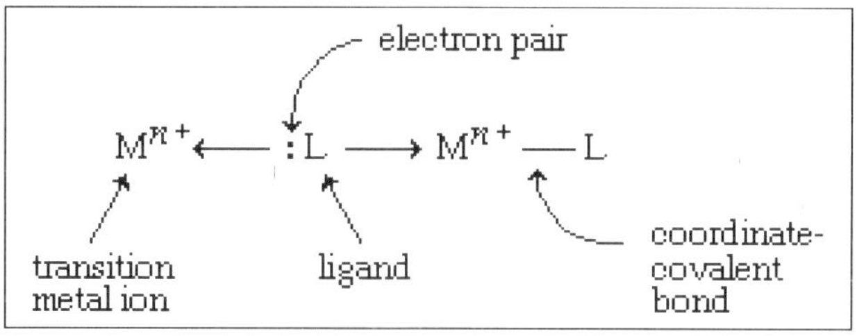

The resulting covalent bond is given a special name because one entity (the ligand) furnishes both of the electrons that are subsequently shared in the bond. An example of a coordination compound is $[Co(NH_3)_6]Cl_3$, which contains the $Co(NH_3)_6^{3+}$ ion, a cobalt ion (Co^{3+}) with six ammonia molecules (NH_3) attached to it, acting as ligands.

In the early days of the science of chemistry, there was no systematic approach to naming compounds. Chemists coined names such as sugar of lead, quicklime, milk of magnesia, Epsom salts, and laughing gas to describe familiar compounds. Such names are called common or trivial names. As chemistry advanced, it became evident that, if common names were used for all known compounds, which number in the millions, great confusion would result. It clearly would be impossible to memorize trivial names for such a large number of compounds. Therefore a systematic nomenclature (naming process) has been developed. There are, however, certain familiar compounds that are always referred to by their common names. The systematic names for H_2O and NH_3, for example, are never used; these vital compounds are known only as water and ammonia, respectively.

The simplest chemical compounds are binary compounds—those consisting of two elements. Different rules apply for the nomenclature of binary ionic compounds and binary molecular (covalent) compounds, and so they will be considered separately.

Binary Compounds

Binary Ionic Compounds

The nomenclature for binary ionic compounds simply entails naming the ions according to the following rules:

1. The positive ion (called a cation) is named first and the negative ion (anion) second.

2. A simple cation (obtained from a single atom) takes its name from its parent element. For example, Li^+ is called lithium in the names of compounds containing this ion. Similarly, Na^+ is called sodium, Mg^{2+} is called magnesium, and so on.

3. A simple anion (obtained from a single atom) is named by taking the root of the parent element's name and adding the suffix -ide. Thus, the F^- ion is called fluoride, Br^- is called bromide, S^{2-} is called sulfide, and so on.

The following examples illustrate the nomenclature rules for binary ionic compounds:

Compound	Ions present	Name
NaCl	Na^+, Cl^-	Sodium chloride
KI	K^+, I^-	Potassium iodide
CaS	Ca^{2+}, S^{2-}	Calcium sulfide
CsBr	Cs^+, Br^-	Cesium bromide
MgO	Mg^{2-}, O^{2-}	Magnesium oxide

In the formulas of ionic compounds, simple ions are represented by the chemical symbol for the element: Cl means Cl^-, Na means Na^+, and so on. When individual ions are shown, however, the charge is always included. Thus, the formula of potassium bromide is given as KBr, but, when the potassium and bromide ions are shown individually, they are written K^+ and Br^-.

When a given metal atom can form more than one type of cation, the charge on the particular cation present must be specified in the name of the compound. For example, lead (Pb) can exist as Pb^{2+} or Pb^{4+} ions in ionic compounds. Also, iron (Fe) can form Fe^{2+} or Fe^{3+} ions, tin (Sn) can form Sn^{2+} or Sn^{4+} ions, gold (Au) can form Au^+ or Au^{3+} ions, and so on. Therefore, the names of binary compounds containing metals such as these must include a Roman numeral to specify the charge on the ion. For example, the compound $FeCl_3$, which contains Fe^{3+}, is named iron(III) chloride. On the other hand, the compound $FeCl_2$, which contains Fe^{2+}, is designated as iron(II) chloride. In each case, the Roman numeral in the name specifies the charge of the metal ion present.

Common simple cations and anions			
Cation	Name	Anion	Name
H^+	Hydrogen	H^-	Hydride
Li^+	Lithium	F^-	Fluoride
Na^+	Sodium	Cl^-	Chloride
K^+	Potassium	Br^-	Bromide
Cs^+	Cesium	I^-	Iodide
Be^{2+}	Beryllium	O^{2-}	Oxide
Mg^{2+}	Magnesium	S^{2-}	Sulfide

Ca^{2+}	Calcium		
Ba^{2+}	Barium		
Al^{3+}	Aluminum		
Ag^{+}	Silver		

An alternative system for naming compounds containing metals that form only two ions is sometimes seen, especially in older literature. The ion with the higher charge has a name ending in -ic, and the one with the lower charge has the suffix -ous. For example, Fe^{3+} is called the ferric ion, and Fe^{2+} is called the ferrous ion. The names for $FeCl_3$ and $FeCl_2$ are then ferric chloride and ferrous chloride, respectively.

Common ions that form multiple cations		
*Mercury(I) ions always occur bound together to form Hg_2^{2+}.		
Ion	Systematic name	Alternate name
Fe^{3+}	Iron(Iii)	Ferric
Fe^{2+}	Iron(Ii)	Ferrous
Cu^{2+}	Copper(Ii)	Cupric
Cu^{+}	Copper(I)	Cuprous
Co^{3+}	Cobalt(Iii)	Cobaltic
Co^{2+}	Cobalt(Ii)	Cobaltous
Sn^{4+}	Tin(Iv)	Stannic
Sn^{2+}	Tin(Ii)	Stannous
Pb^{4+}	Lead(Iv)	Plumbic
Pb^{2+}	Lead(Ii)	Plumbous
Hg^{2+}	Mercury(Ii)	Mercuric
Hg_2^{2+}(*)	Mercury(I)	Mercurous

Binary Molecular (Covalent) Compounds

Binary molecular (covalent) compounds are formed as the result of a reaction between two nonmetals. Although there are no ions in these compounds, they are named in a similar manner to binary ionic compounds. The nomenclature of binary covalent compounds follows these rules:

1. The first element in the formula is given first, using the element's full name.
2. The second element is named as if it were an anion.
3. Prefixes are used to denote the numbers of atoms present. If the first element exists as a single atom, the prefix mono is omitted. For example, CO is called carbon monoxide rather than monocarbon monoxide.

These examples show how the rules are applied for the covalent compounds formed by nitrogen and oxygen:

Compound	Systematic name	Common name
N_2O	Dinitrogen monoxide	Nitrous oxide (laughing gas)
NO	Nitrogen monoxide	Nitric oxide
NO_2	Nitrogen dioxide	
N_2O_3	Dinitrogen trioxide	
N_2O_4	Dinitrogen tetroxide	
N_2O_5	Dinitrogen pentoxide	

To avoid awkward pronunciations, the final o or a of the prefix is often dropped when the element name begins with a vowel. For example, N_2O_4 is referred to as dinitrogen tetroxide, not dinitrogen tetraoxide, and CO is called carbon monoxide, not carbon monooxide.

Prefixes used in chemical nomenclature	
Prefix	Number of atoms
mono-	1
di-	2
tri-	3
tetra-	4
penta-	5
hexa-	6
hepta-	7
octa-	8

Nonbinary Compounds

Ionic Compounds Containing Polyatomic Ions

A special type of ionic compound is exemplified by ammonium nitrate (NH_4NO_3), which contains two polyatomic ions, NH_4^+ and NO_3^-. As the name suggests, a polyatomic ion is a charged entity composed of several atoms bound together. Polyatomic ions have special names that are used in the nomenclature of the compounds containing them.

Common polyatomic ions			
*Bisulfate and **bicarbonate are widely used common names for hydrogen sulfate and hydrogen carbonate, respectively.			
Ion	Name	Ion	Name
NH_4^+	Ammonium	CO_3^{2-}	Carbonate
NO_2^-	Nitrite	HCO_3^-	Hydrogen carbonate**

NO_3^-	Nitrate	ClO^-	Hypochlorite
SO_3^{2-}	Sulfite	ClO_2^-	Chlorite
SO_4^{2-}	Sulfate	ClO_3^-	Chlorate
HSO_4^-	Hydrogen Sulfate*	ClO_4^-	Perchlorate
OH^-	Hydroxide	$C_2H_3O_2^-$	Acetate
CN^-	Cyanide	MnO_4^-	Permanganate
PO_4^{3-}	Phosphate	$Cr_2O_7^{2-}$	Dichromate
HPO_4^{2-}	Hydrogen Phosphate	CrO_4^{2-}	Chromate
$H_2PO_4^-$	Dihydrogen Phosphate	O_2^{2-}	Peroxide

Several series of polyatomic anions exist that contain an atom of a given element in combination with different numbers of oxygen atoms. Such anions are called oxy anions. When the series contains only two members, the name of the ion with fewer oxygen atoms ends in -ite, and the name of the other ion ends in -ate. For example, SO_3^{2-} is called sulfite and SO_4^{2-} is called sulfate. In those cases where more than two oxy anions constitute the series, hypo- (less than) and per- (more than) are used as prefixes to name the members of the series with the smallest and the largest number of oxygen atoms, respectively. The chlorine-containing oxy anions provide an example:

ClO^-	Hypochlorite
ClO_2^-	Chlorite
ClO_3^-	Chlorate
ClO_4^-	Perchlorate

Naming ionic compounds that contain polyatomic ions is similar to naming binary ionic compounds. For example, the compound NaOH is called sodium hydroxide, because it contains the Na^+ (sodium) cation and the OH^- (hydroxide) anion. As in binary ionic compounds, when a metal that can form multiple cations is present, a Roman numeral is required to specify the charge on the cation. For example, the compound $FeSO_4$ is called iron(II) sulfate, because it contains Fe^{2+}.

Acids

An acid can be thought of as a molecule containing at least one hydrogen cation (H^+) attached to an anion. The nomenclature of acids depends on whether the anion contains oxygen. If the anion does not contain oxygen, the acid is named with the prefix hydro- and the suffix -ic. For example, HCl dissolved in water is called hydrochloric acid. Likewise, HCN and H_2S dissolved in water are called hydrocyanic and hydrosulfuric acids, respectively.

If the anion of the acid contains oxygen, the name is formed by adding the suffix -ic or -ous to the root name of the anion. If the anion name ends in -ate, the -ate is replaced by -ic (or sometimes -ric). For example, H_2SO_4 contains the sulfate anion (SO_4^{2-}) and is called sulfuric acid; H_3PO_4 contains the phosphate anion (PO_4^{3-}) and is called phosphoric acid; and $HC_2H_3O_2$, which contains the acetate ion ($C_2H_3O_2^-$), is called acetic acid. For anions with an -ite ending, the -ite is replaced by -ous in naming the acid. For example, H_2SO_3, which contains sulfite (SO_3^{2-}), is called sulfurous acid; and HNO_2, which contains nitrite (NO_2^-), is named nitrous acid. The acids of the oxy anions of chlorine are used here to illustrate the rules for naming acids with oxygen-containing cations.

Names of less common acids	
*Often called boric acid. **Often called phosphoric acid.	
Formula	Name
H_3BO_3	Orthoboric acid*
H_2CO_3	Carbonic acid
H_3PO_4	Orthophosphoric acid**
$H_4P_2O_7$	Pyrophosphoric acid
$H_5P_3O_{10}$	Triphosphoric acid
$(HPO_3)n$	Metaphosphoric acid
$(HPO_3)_3$	Trimetaphosphoric acid
H_3PO_3	Phosphorous acid
H_3PO_2	Hypophosphorous acid
H_2SO_5	Peroxosulfuric acid
$H_2S_2O_6$	Dithionic acid
$H_2S_2O_3$	Thiosulfuric acid
$HMnO_4$	Permanganic acid

Names of common acids	
HF	Hydrofluoric acid
HCl	Hydrochloric acid
HBr	Hydrobromic acid
HI	Hydroiodic acid
HCN	Hydrocyanic acid
H_2S	Hydrosulfuric acid
HNO_3	Nitric acid
HNO_2	Nitrous acid
H_2SO_4	Sulfuric acid
H_2SO_3	Sulfurous acid
$HC_2H_3O_2$	Acetic acid

Acid formula	Anion	Name
$HClO_4$	Perchlorate	Perchloric acid
$HClO_3$	Chlorate	Chloric acid
$HClO_2$	Chlorite	Chlorous acid
HClO	Hypochlorite	Hypochlorous acid

Compounds with Complex Ions

A coordination compound is composed of one or more complex structural units, each of which has a central atom bound directly to a surrounding set of groups called ligands. The nomenclature of coordination compounds is based on these structural relationships.

Organic Compounds

In general, organic compounds are substances that contain carbon (C), and carbon atoms provide the key structural framework that generates the vast diversity of organic compounds. All things on the Earth (and most likely elsewhere in the universe) that can be described as living have a crucial dependence on organic compounds. Foodstuffs—namely, fats, proteins, and carbohydrates—are organic compounds, as are such vital substances as hemoglobin, chlorophyll, enzymes, hormones, and vitamins. Other materials that add to the comfort, health, or convenience of humans are composed of organic compounds, including clothing made of cotton, wool, silk, and synthetic fibres; common fuels, such as wood, coal, petroleum, and natural gas; components of protective coatings, such as varnishes, paints, lacquers, and enamels; antibiotics and synthetic drugs; natural and synthetic rubber; dyes; plastics; and pesticides.

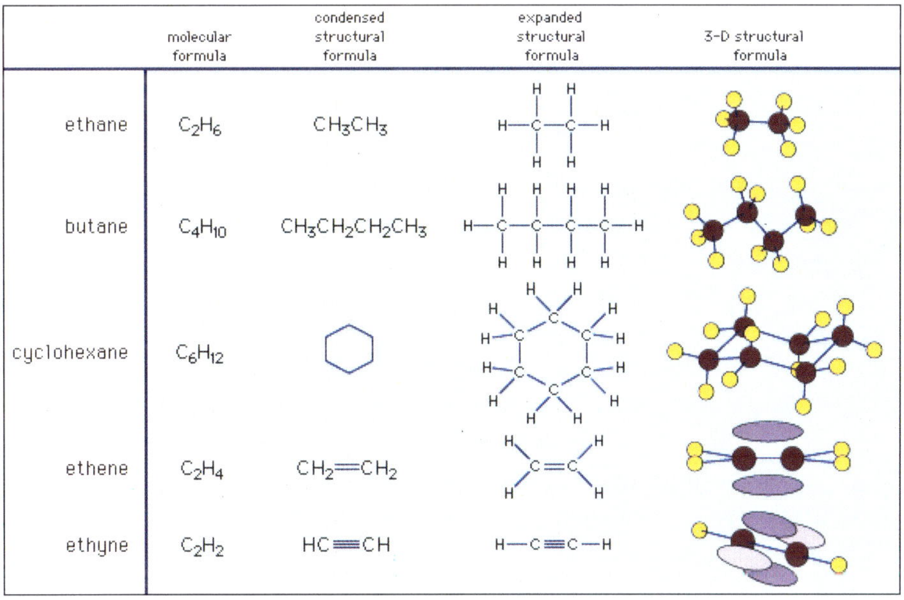

The structures of organic compounds can be depicted in condensed, expanded, and three-dimensional structural formulas.

Carbon Bonding

The carbon atom is unique among elements in its tendency to form extensive networks of covalent bonds not only with other elements but also with itself. Because of its position midway in the second horizontal row of the periodic table, carbon is neither an electropositive nor an electronegative element; it therefore is more likely to share electrons than to gain or lose them. Moreover, of all the elements in the second row, carbon has the maximum number of outer shell electrons (four) capable of forming covalent bonds. (Other elements, such as phosphorus [P] and cobalt [Co], are able to form five and six covalent bonds, respectively, with other elements, but they lack carbon's ability to bond indefinitely with itself.) When fully bonded to other atoms, the four bonds of the carbon atom are directed to the corners of a tetrahedron and make angles of about 109.5° with each other. The result is that not only can carbon atoms combine with one another indefinitely to give compounds of extremely high molecular weight, but the molecules formed can exist in an infinite variety of three-dimensional structures. The possibilities for diversity are increased by the presence of atoms other than carbon in organic compounds, especially hydrogen (H), oxygen (O), nitrogen (N), halogens (fluorine [F], chlorine [Cl], bromine [Br], and iodine [I]), and sulfur (S). It is the enormous potential for variation in chemical properties that has made organic compounds essential to life on Earth.

The structures of organic compounds commonly are represented by simplified structural formulas, which show not only the kinds and numbers of atoms present in the molecule but also the way in which the atoms are linked by the covalent bonds—information that is not given by simple molecular formulas, which specify only the number and type of atoms contained in a molecule. (With most inorganic compounds, the use of structural formulas is not necessary, because only a few atoms are involved and only a single arrangement of the atoms is possible.) In the structural formulas of organic compounds, short lines are used to represent the covalent bonds. Atoms of the individual elements are represented by their chemical symbols, as in molecular formulas.

Structural formulas vary widely in the amount of three-dimensional information they convey, and the type of structural formula used for any one molecule depends on the nature of the information the formula is meant to display. The different levels of sophistication can be illustrated by considering some of the least complex organic compounds, the hydrocarbons. The gas ethane, for example, has the molecular formula C_2H_6. The simplest structural formula, drawn either in a condensed or in an expanded version, reveals that ethane consists of two carbon atoms bonded to one another, each carbon atom bearing three hydrogen atoms. Such a two-dimensional representation correctly shows the bonding arrangement in ethane, but it does not convey any information about its three-dimensional architecture. A more sophisticated structural formula can be drawn to better represent the three-dimensional structure of the molecule. Such a structural formula correctly shows the tetrahedral orientation of the four atoms (one carbon and three hydrogens) bonded to each carbon, and the specific architecture of the molecule.

Larger organic molecules are formed by the addition of more carbon atoms. Butane, for example, is a gaseous hydrocarbon with the molecular formula C_4H_{10}, and it exists as a chain of four carbon atoms with 10 attached hydrogen atoms. As carbon atoms are added to a molecular framework, the carbon chain can develop branches or form cyclic structures. A very common ring structure contains six carbon atoms in a ring, each bonded in a tetrahedral arrangement, as in the hydrocarbon cyclohexane, C_6H_{12}. Such ring structures are often very simply represented as regular polygons in which each apex represents a carbon atom, and the hydrogen atoms that complete the bonding requirements of the carbon atoms are not shown. The polygon convention for cyclic structures reveals concisely the bonding arrangement of the molecule but does not explicitly convey information about the actual three-dimensional architecture. It should be noted that the polygon is only a two-dimensional symbol for the three-dimensional molecule.

Under certain bonding conditions, adjacent atoms will form multiple bonds with each other. A double bond is formed when two atoms use two electron pairs to form two covalent bonds; a triple bond results when two atoms share three electron pairs to form three covalent bonds. Multiple bonds have special structural and electronic features that generate interesting chemical properties. The six atoms involved in a double bond (as in ethene, C_2H_4) lie in a single plane, with regions above and below the plane occupied by the electrons of the second covalent bond. Atoms in a triple bond (as in acetylene, or ethyne, C_2H_2) lie in a straight line, with four regions beside the bond axis occupied by electrons of the second and third covalent bonds.

Functional Groups

Chemists observed early in the study of organic compounds that certain groups of atoms and associated bonds, known as functional groups, confer specific reactivity patterns on the molecules of which they are a part. Although the properties of each of the several million organic molecules whose structure is known are unique in some way, all molecules that contain the same functional group have a similar pattern of reactivity at the functional group site. Thus, functional groups are a key organizing feature of organic chemistry. By focusing on the functional groups present in a molecule (most molecules have more than one functional group), several of the reactions that the molecule will undergo can be predicted and understood.

Because carbon-to-carbon and carbon-to-hydrogen bonds are extremely strong and the charge of the electrons in these covalent bonds is spread more or less evenly over the bonded atoms, hydrocarbons that contain only single bonds of these two types are not very reactive. The reactivity of a molecule increases if it contains one or more weak bonds or bonds that have an unequal distribution of electrons between the two atoms. If the two electrons of a covalent bond are, for one reason or another, drawn more closely to one of the bonded atoms, that atom will develop a partial negative charge and the atom to which it is bonded will develop a partial positive charge. A covalent bond in which the electron pair linking the atoms is shared unequally is known as a polar bond. Polar bonds, and any other bonds that have unique electronic properties, confer the potential for chemical reaction on the molecule in which they are present. This is because, for every reaction, one or more bonds of a molecule must be broken and new bonds formed. The presence of a partial negative charge (a region of high electron density) will draw to itself other atoms or groups of atoms that are deficient in electron density. This initiates the process of bond breaking that is a prerequisite for a chemical reaction. For these reasons, molecules with regions of increased or decreased electron density are especially important for chemical change.

Common functional groups

functional group	functional group name	generalized molecular formula	organic classification	examples
>C=C<	double bond	$R_2C=CR_2$	alkenes	$CH_2=CH_2$ ethene (ethylene); $CH_3CH=CH_2$ propene (propylene); $CH_2=CH-CH=CH_2$ butadiene
—C≡C—	triple bond	R—C≡C—R	alkynes	HC≡CH ethyne (acetylene); $HC≡CCH_2CH_3$ butyne
	benzene ring	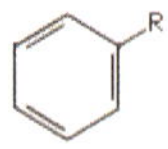	aromatics	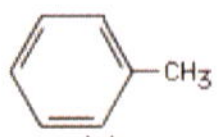 toluene; 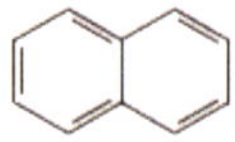 naphthalene
—OH	hydroxyl	R—OH	alcohols	CH_3CH_2OH ethanol; 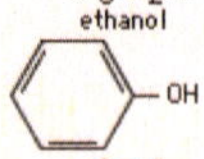 phenol
—OR	ethoxy	R—OR	ethers	CH_3OCH_3 dimethyl ether
—C(=O)H	aldehyde	R—CHO	aldehydes	CH_3CHO ethanal (acetaldehyde)
>C=O	carbonyl	R—C(=O)—R	ketones	$CH_3C(=O)CH_2CH_3$ 2-butanone (ethyl methyl ketone)
—C(=O)OH	carboxyl	R—COOH	carboxylic acids	CH_3COOH ethanoic acid (acetic acid)
$—NR_2$	amino	$R—NR_2$	amines	CH_3NHCH_3 dimethylamine
—SH	mercapto	R—SH	thiols	CH_3SH methane thiol
—F, Cl, Br, I	fluoro, chloro, bromo, iodo	R—X	halides	$H_2C=CHCl$ chloroethene (vinyl chloride)

There are two major bonding features that generate the reactive sites of functional groups. The first, already mentioned, is the presence of multiple bonds. Both double and triple bonds have regions of high electron density lying outside the atom-to-atom bond axis. Double and triple bonds are known as functional groups, a term that is used to identify atoms or groups of atoms within a molecule that are sites of comparatively high reactivity. A second type of reactive site results when an atom other than carbon or hydrogen (termed a heteroatom) is bonded to carbon. All heteroatoms have a greater or lesser attraction for electrons than does carbon. Thus, each bond between a carbon and a heteroatom is polar, and the degree of polarity depends on the difference between the electron-attracting properties of the two atoms. The most important atomic groupings that contain such reactive polar bonds are also able to generate functional groups.

To emphasize the generality of reactions between molecules that contain the same functional group, chemists often represent the less reactive portions of a molecule by the symbol R. Thus, all molecules that contain a double bond, however complicated, can be represented by the general formula for an alkene—i.e.,

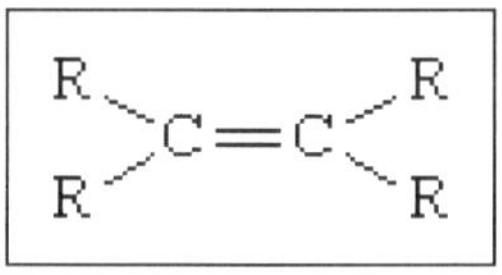

This type of formula suggests that the molecule will undergo those reactions that are common to double bonds and that the reaction will occur at the double bond. The rest of the molecule, represented by the four R groups, will remain unchanged by the reaction occurring at the functional group site.

Molecules with more than one functional group, called polyfunctional, may have more complicated properties that result from the identity—and interconnectedness—of the multiple functional groups. Many natural products contain several functional groups located at specific sites within a large, complicated, three-dimensional structure.

Alkanes

Alkanes are compounds that consist entirely of atoms of carbon and hydrogen (a class of substances known as hydrocarbons) joined to one another by single bonds. The shared electron pair in each of these single bonds occupies space directly between the two atoms; the bond generated by this shared pair is known as a sigma (σ) bond. Both carbon-carbon and carbon-hydrogen sigma bonds are single strong, nonpolar covalent bonds that are normally the least reactive bonds in organic molecules. Alkane sequences form the inert framework of most organic compounds. For this reason, alkanes are not formally considered a functional group. When a hydrocarbon chain is connected as a substituent to a more fundamental structural unit, it is termed an alkyl group. The simplest examples of alkanes are methane (CH_4; the principal constituent of natural gas), ethane (C_2H_6), propane (C_3H_8; widely used as a barbecue fuel), and butane (C_4H_{10};

the liquid fuel in pocket lighters). Hydrocarbon chains commonly occur in cyclic forms, or rings; the most common example is cyclohexane (C_6H_{12}).

Alkenes

Organic compounds are termed alkenes if they contain a carbon-carbon double bond. The shared electron pair of one of the bonds is a σ bond. The second pair of electrons occupies space on both sides of the σ bond; this shared pair constitutes a pi (π) bond. A π bond forms a region of increased electron density because the electron pair is more distant from the positively charged carbon nuclei than is the electron pair of the σ bond. Even though a carbon-carbon double bond is very strong, a π bond will draw to itself atoms or atomic groupings that are electron-deficient, thereby initiating a process of bond-breaking that can lead to rupture of the π bond and formation of new σ bonds. A simple example of an alkene reaction, which illustrates the way in which the electronic properties of a functional group determine its reactivity, is the addition of molecular hydrogen to form alkanes, which contain only σ bonds.

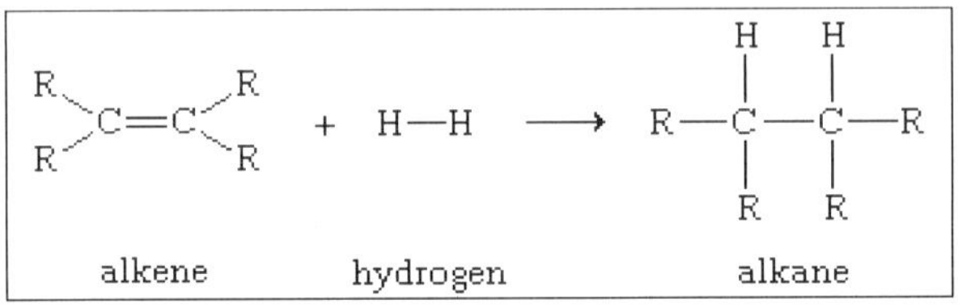

Such reactions, in which the π bond of an alkene reacts to form two new σ bonds, are energetically favourable because the new bonds formed (two carbon-hydrogen σ bonds) are stronger than the bonds broken (one carbon-carbon π bond and one hydrogen-hydrogen σ bond). Because the addition of atoms to the π bond of alkenes to form new σ bonds is a general and characteristic reaction of alkenes, alkenes are said to be unsaturated. Alkanes, which cannot be transformed by addition reactions into molecules with a greater number of σ bonds, are said to be saturated.

The alkene functional group is an important one in chemistry and is widespread in nature. Some common examples include ethylene (used to make polyethylene), 2-methyl-1,3-butadieneisoprene (used to make rubber), and vitamin A (essential for vision).

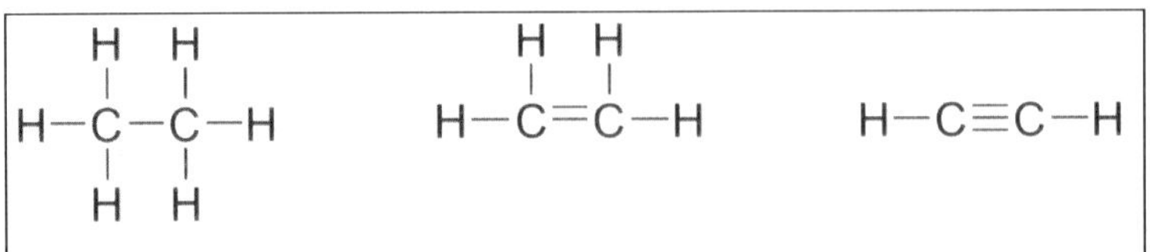

eth**ane** (an alk**ane**) — eth**ene** (an alk**ene**) — eth**yne** (an alk**yne**)

For ethene, both the carbon atoms of an alkene and the four atoms connected to the double bond lie in a single plane.

Alkynes

Molecules that contain a triple bond between two carbon atoms are known as alkynes. The triple bond is made up of one σ bond and two π bonds. As in alkenes, the π bonds constitute regions of increased electron density lying parallel to the carbon-carbon bond axis. Carbon-carbon triple bonds are very strong bonds, but reactions do occur that break the π bonds to form stronger σ bonds.

The most common example of an alkyne is ethyne (also known as acetylene), used as a fuel for oxyacetylene torches in welding applications. Alkynes are not abundant in nature, but the fungicide capillan contains two alkyne functional groups.

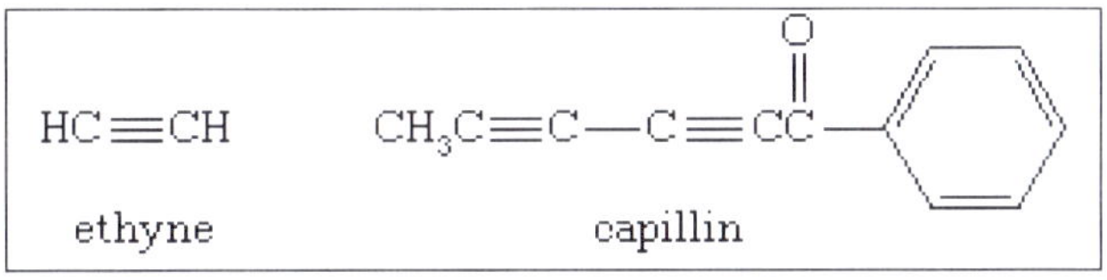

Aromatic Hydrocarbons (Arenes)

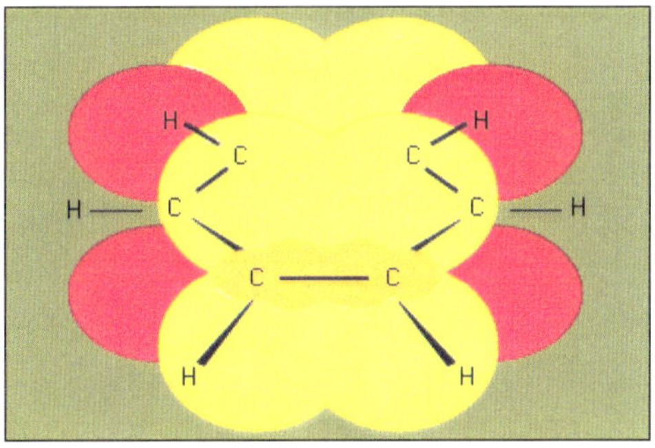

Benzene is the smallest of the organic aromatic hydrocarbons. It contains sigma bonds (represented by lines) and regions of high-pi electron density, formed by the overlapping of p orbitals (represented by the dark yellow shaded area) of adjacent carbon atoms, which give benzene its characteristic planar structure.

A distinctive set of physical and chemical properties is imparted to molecules that contain a functional group composed of three pairs of doubly bonded atoms (usually all carbon atoms) bonded together in the shape of a regular planar (flat) hexagon. The hexagonal ring is usually drawn with an alternating sequence of single and double bonds. The molecule benzene, C_6H_6, first discovered by English physicist and chemist Michael Faraday in 1825, is the smallest molecule that can contain this functional group, and arenes contain one or more benzene (or structurally similar) rings. Because benzene and many larger arenes have a strong odour, they have long been known as aromatic hydrocarbons. Benzene, and all the larger arenes, have a characteristic planar structure forced on them by the electronic requirements of the six (or more) pi electrons. When named as substituents on other structural units, the aromatic units are called aryl substituents.

Naphthalene, the active component of mothballs, contains two fused benzene rings. Benzoapyrene, an aromatic hydrocarbon produced in small amounts by the combustion of organic substances, contains five fused benzene rings. Like several other polycyclic aromatic hydrocarbons, it is carcinogenic. Aromatic compounds are widely distributed in nature. Benzaldehyde, anisole, and vanillin, for example, have pleasant aromas.

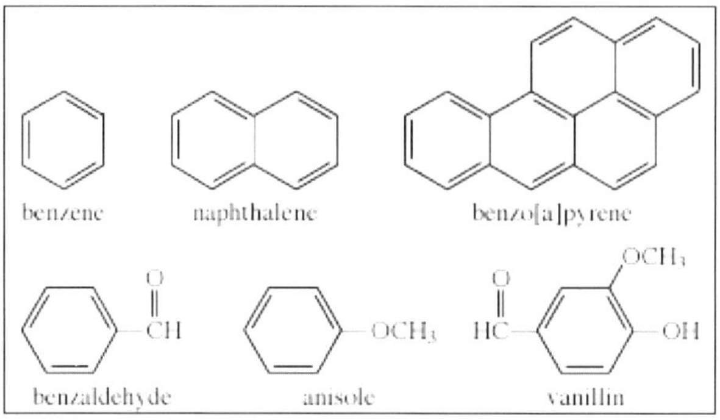

Alcohols and Phenols

An oxygen atom normally forms two σ bonds with other atoms; the water molecule, H_2O, is the simplest and most common example. If one hydrogen atom is removed from a water molecule, a hydroxyl functional group (−OH) is generated. When a hydroxyl group is joined to an alkane framework, an alcohol such as ethanol, is produced.

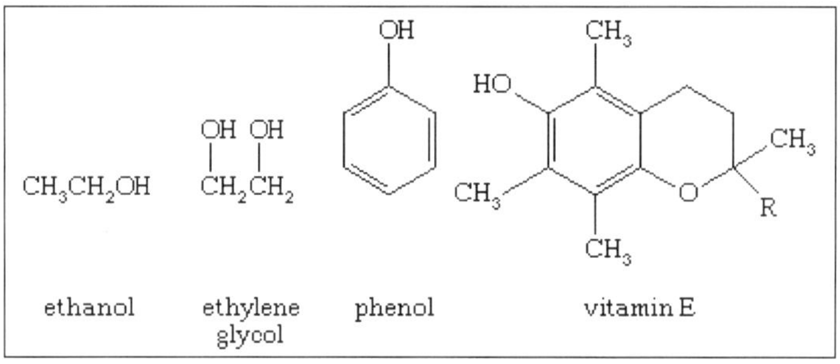

When the hydroxyl group is joined to an aryl ring, a phenol results. Both alcohols and phenols are widespread in nature, with alcohols being especially ubiquitous. The hydroxyl group of alcohols and phenols is responsible for an interesting variety of physical and chemical properties. The biochemical action of vitamin E, for example, depends largely on the reactivity of the phenol functional group.

An oxygen atom is much more electronegative than carbon or hydrogen atoms, so both carbon-oxygen and hydrogen-oxygen bonds are polar. The oxygen atom is slightly negatively charged, and the carbon and hydrogen atoms are slightly positively charged. The polar bonds of the hydroxyl group are responsible for the major reaction characteristics of alcohols and phenols. In general, these reactions are initiated by reaction of electron-deficient groups with the negatively charged oxygen atom or by reaction of electron-rich groups with the positively charged atoms—namely, carbon or hydrogen—bonded to oxygen.

Ethers and Epoxides

An organic molecule in which an oxygen atom is bonded to two carbon atoms through two sigma bonds is known as an ether. Ether molecules occur widely in nature. Diethyl ether was once widely used as an anesthetic. An aromatic ether known as Nerolin II (2-ethoxynaphthalene) is used in perfumes to impart the scent of orange blossoms. Cyclic ethers, such as tetrahydrofuran, are commonly used as organic solvents. Although ethers contain two polar carbon-oxygen bonds, they are much less reactive than alcohols or phenols.

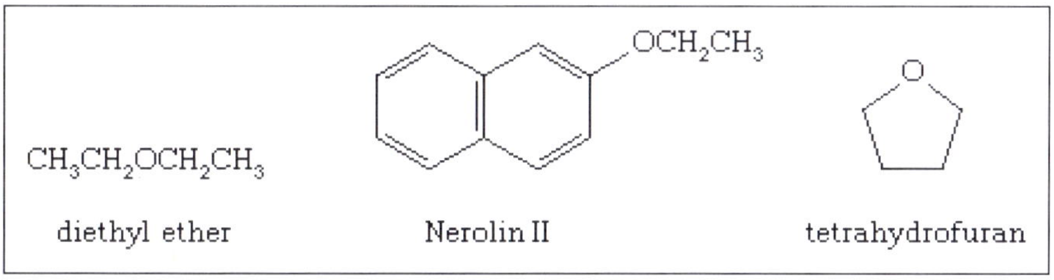

Thiols

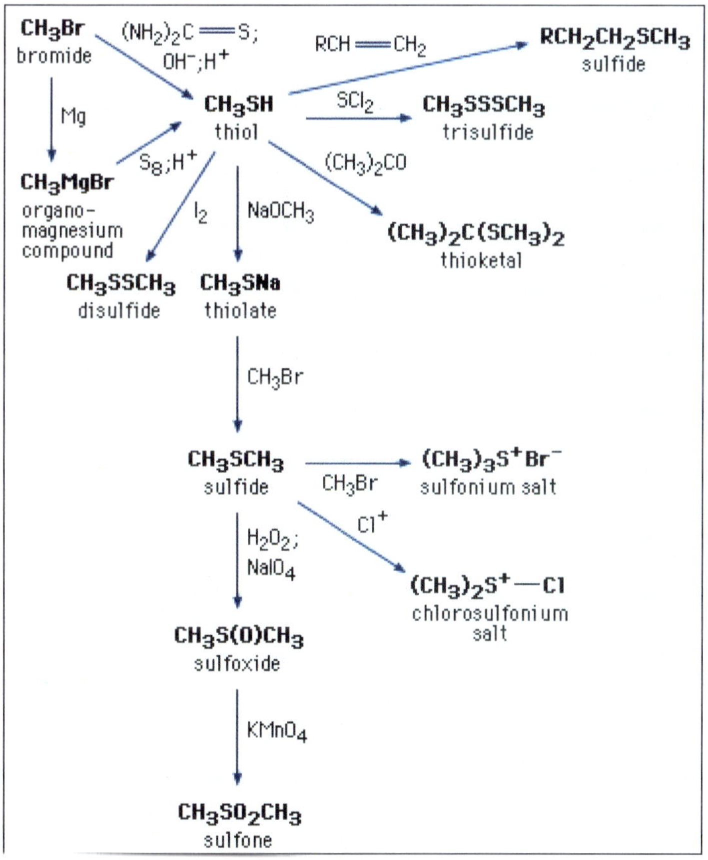

Thiols are organic functional groups, similar in structure to alcohols but containing sulfur atoms in place of oxygen atoms. Thiols react to form sulfide compounds and other sulfur-containing compounds.

A thiol is structurally similar to an alcohol but contains a sulfur atom in place of the oxygen atom normally found in an alcohol. The outstanding feature of thiols is their foul smell. The simplest thiol is hydrogen sulfide, H_2S, the sulfur analog of water. It can be detected by the human nose at a concentration of a few parts per billion and is readily identifiable as having a rotten-egg odour. Ethanethiol is added in trace amounts to natural gas to give it a detectable odour, and striped skunks deter predators by releasing a liquid spray containing 3-methyl-1-butanethiol. When present as a substituent on another structural unit, the SH group is commonly termed mercapto, as in 2-mercaptoethanol.

Amines

Amines are functional group compounds that contain at least one nitrogen atom bonded to hydrogen atoms or to alkyl or aryl groups. If the substituents (other than hydrogen atoms) are alkyl groups, the resulting compounds are termed alkyl amines. If one or more substituents is an aryl group, the compounds are termed aryl amines. Amines are commonly categorized as primary, secondary, or tertiary, depending on whether the nitrogen atom is bonded to one, two, or three alkyl or aryl groups, respectively. The nitrogen atom is bonded to its hydrogen atoms and alkyl groups by sigma (σ) bonds, but the nitrogen atom also bears a nonbonded electron pair. The three σ bonds and nonbonded electron pair are oriented around the nitrogen atom in a distorted tetrahedral geometry.

In some compounds, the nonbonded electron pair on the nitrogen atom is replaced by a fourth σ bond to a hydrogen atom or to an alkyl or aryl group. The resulting compound, called a quaternary ammonium salt, has a positive charge on the nitrogen atom and a tetrahedral arrangement of groups around the nitrogen atom. Amines are very common organic molecules, and many are physiologically active. Amphetamine, for example, is a central nervous system stimulant and acts as an antidepressant. Amines are particularly valuable because of their ability to act as bases, a property that is a consequence of the ability of amines to accept hydrogen atoms from acidic molecules.

Halides

Halides, or organohalides, are compounds that contain a halogen atom (fluorine, chlorine, bromine, or iodine) bonded to a carbon atom by a polar bond. The slightly positive charge that exists on the carbon atom in carbon-halogen bonds is the source of reactivity of halides. A wide variety of organohalides have been discovered in marine organisms, and several simple halide compounds have important commercial applications. Chloroethane (ethyl chloride) is a volatile liquid that is used as a topical anesthetic. Chloroethene (vinyl chloride) is the monomeric building block for polyvinyl chloride (PVC), and the mixed organohalide halothane is an inhalation anesthetic. The compound epibatidine, isolated from glands on the back of an Ecuadorian poison frog, has been found to be an especially potent painkiller.

Aldehydes and Ketones

When an oxygen atom forms a double bond to a carbon atom, a carbonyl functional group is obtained. The carbon atom of a carbonyl group is bonded to two other atoms in addition to the oxygen atom. A wide range of functional groups are produced by the presence of different atomic groupings on the carbon of the carbonyl group. Two of the most important are aldehydes and ketones. In a ketone, both atoms bonded to the carbonyl carbon are other carbon atoms, and, in an aldehyde, at least one atom on the carbonyl carbon is a hydrogen. Similar to the double bond of alkenes, the carbon-oxygen double bond is made up of a σ bond, whose electron pair lies between the bonded atoms, and a π bond, whose electron pair occupies space on both sides of the σ bond.

Many aldehydes and ketones have pleasant, fruity aromas, and these compounds are frequently responsible for the flavour and smell of fruits and vegetables. A 40 percent solution of formaldehyde in water is formalin, a liquid used for preserving biological specimens. Benzaldehyde is an aromatic aldehyde and imparts much of the aroma to cherries and almonds. Butanedione, a ketone with two carbonyl groups, is partially responsible for the odour of cheeses. Civetone, a large cyclic ketone, is secreted by the civet cat and is a key component of many expensive perfumes.

The carbonyl group has a wide variety of reaction pathways open to it. Because of its π bond, the carbonyl group undergoes addition reactions similar to those that occur with alkenes but with a few important differences. Whereas carbon-carbon double bonds are nonpolar, carbon-oxygen double bonds are polar. Species that add to a carbonyl group to form new σ bonds react in such a way that electrophilic (electron-seeking) groups attack the oxygen atom and nucleophilic groups (those seeking positively charged centres) attack the carbon atom. Furthermore, addition to a carbonyl group results in the breaking of a strong π bond. The energy relationships of carbonyl addition reactions are consequently very different from those of alkene addition reactions. Other reaction possibilities of carbonyl compounds depend on the nature of the atomic groupings, termed substituents, attached to the carbonyl carbon. When both substituents are unreactive alkane fragments, as in ketones, there are few reactions other than carbonyl additions. When one of the substituents is not an alkane fragment, different possibilities emerge. In aldehydes, the carbonyl carbon is bonded to a hydrogen atom, and reactions that involve this hydrogen atom distinguish the reactions of aldehydes from those of ketones.

Carboxylic Acids

The conjunction of a carbonyl and a hydroxyl group forms a functional group known as a carboxyl group.

The hydrogen of a carboxyl group can be removed (to form a negatively charged carboxylate ion), and thus molecules containing the carboxyl group have acidic properties and are generally known as carboxylic acids. Vinegar is a 5 percent solution of acetic acid in water, and its sharp acidic taste is due to the carboxylic acid present. Lactic acid provides much of the sour taste of pickles and sauerkraut and is produced by contracting muscles. Citric acid is a major flavour component of citrus fruits, such as lemons, grapefruits, and oranges. Ibuprofen, an effective analgesic and anti-inflammatory agent, contains a carboxyl group.

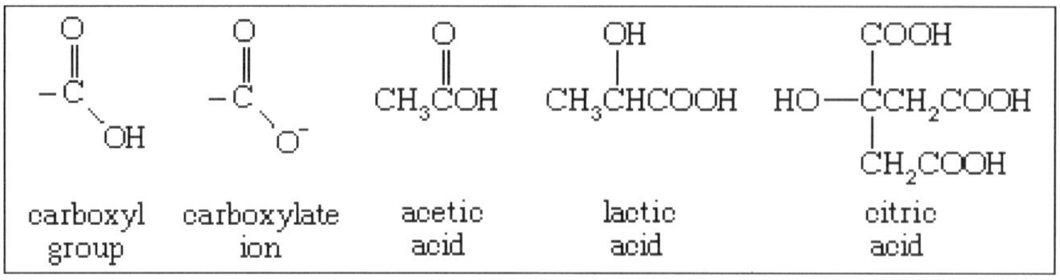

The structural unit containing an alkyl group bonded to a carbonyl group is known as an acyl group. A family of functional groups, known as carboxylic acid derivatives, contains the acyl group bonded to different substituents.

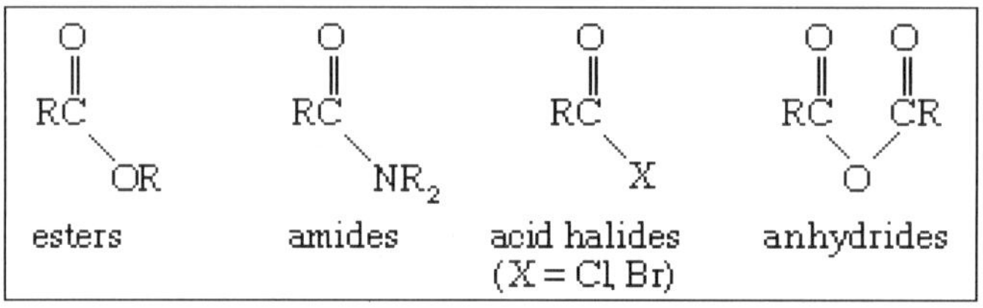

Esters have an alkoxy (OR) fragment attached to the acyl group; amides have attached amino groups ($-NR_2$); acyl halides have an attached chlorine or bromine atom; and anhydrides have an attached carboxyl group. Each type of acid derivative has a set of characteristic reactions that qualifies it as a unique functional group, but all acid derivatives can be readily converted to a carboxylic acid under appropriate reaction conditions. Many simple esters are responsible for the pleasant odours of fruits and flowers. Methyl butanoate, for example, is present in pineapples. Urea, the major organic constituent of urine and a widely used fertilizer, is a double amide of carbonic acid. Acyl chlorides and anhydrides are the most reactive carboxylic acid derivatives and are useful chemical reagents, although they are not important functional groups in natural substances.

Polyfunctional Compounds

Although each of the functional groups introduced above has a characteristic set of favoured reactions, it is not always possible to predict the properties of organic compounds that contain several different functional groups. In polyfunctional organic compounds, the functional groups often interact with one another to impart unique reactivity patterns to the compounds. As chemistry evolves as a science, it becomes possible to understand more of the behaviour of complex molecules, and chemists are able to design laboratory syntheses of increasingly complicated molecules, basing the synthetic plan upon the reactivity trends of functional groups.

Chemical Synthesis

Chemical synthesis is concerned with the construction of complex chemical compounds from simpler ones. A synthesis usually is undertaken for one of three reasons. The first reason is to meet an industrial demand for a product. For example, ammonia is synthesized from nitrogen and hydrogen and is used to make, among other things, ammonium sulfate, employed as a fertilizer; vinyl chloride is made from ethylene and is used in the production of polyvinyl chloride (PVC) plastic. In general, a vast range of chemical compounds are synthesized for applications as fibres and plastics, pharmaceuticals, dyestuffs, herbicides, insecticides, and other products.

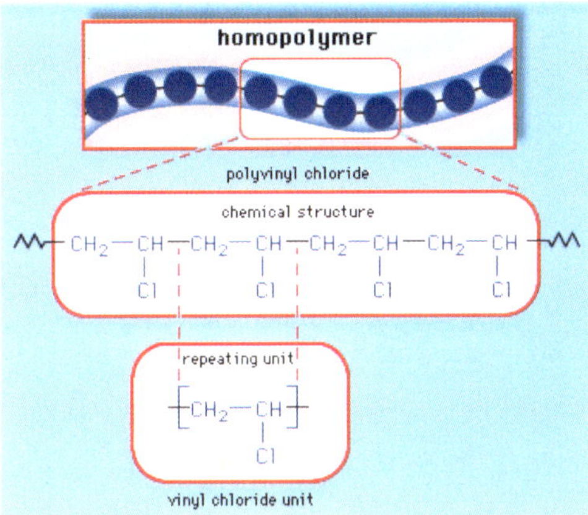

Industrial polymers are synthesized from simple compounds joined together to form long chains. For example, polyvinyl chloride is an industrial homopolymer synthesized from repeating units of vinyl chloride.

Second, an enormous number of compounds of considerable molecular complexity occur naturally, in both living organisms and their degradation products; examples are proteins (in animals) and alkaloids (alkaline materials found in plants). The syntheses of these natural products have usually been undertaken in the context of the determination of the structures of the compounds; if a material is deduced to have a particular structure on the basis of its chemical reactions and physical properties, then the discovery that a compound synthesized by an unambiguous method for this structure is identical to the natural product provides confirmation of the validity of the assigned structure.

Third, a synthesis may be carried out to obtain a compound of specific structure that does not occur naturally and has not previously been made. This type of synthesis is performed in order to examine the properties of the compound and thereby test theories of chemical structure and reactivity.

Approach to Synthesis

The range of compounds that are capable of being synthesized is essentially limitless. In practice, the synthesis of a preselected compound is made possible by particular functional groups undergoing transformations that, while they are dependent on the conditions applied to the compound, are largely independent of the structure of the remaining part of the molecule. Thus, the combination of knowledge of the structure of the compound to be synthesized and knowledge of the general types of transformation that compounds undergo enables a synthesis to be planned. The general approach, cut to its barest essentials, is to examine the structure of the desired end product—for example, Z—and to deduce the structure of some (slightly simpler) compound—for example, Y—that should be capable of transformation into Z by a reaction of known type. A possible precursor of Y is sought in similar manner, and in this way the chain of compounds is extended until a compound, A, is reached that is available for the work; the necessary transformations, beginning with A and ending with Z, are then carried out. Most individual steps in the sequence result in a change in only one bond; some result in changes in two bonds at a time, but it is unusual for more extensive changes to occur.

Evaluation of a Synthetic Method

Three factors must be borne in mind when evaluating a particular synthetic plan. The first is cost—of far greater importance in industrial, large-scale synthesis than in laboratory work in which a particular synthesis may be carried out only once, as in the total synthesis of a naturally occurring compound, and which in any case is likely to be on a relatively small scale. The environmental impact of chemical syntheses has become an important consideration. Syntheses or processes that have a benign environmental impact, whether by use of safe and commonly available reagents or by minimization of environmentally harmful waste products, have become an essential feature of so-called "green chemistry."

Second, the yield in each step must be considered. A step in a synthesis may give a very low yield of the desired product. For example, a proportion of the reactant may be converted into a different product by an alternative process that competes with the desired one; some of the product may undergo a subsequent reaction; or some of the product may be lost in the separation processes required for its isolation in a pure state. The yield is usually defined, on a percentage basis, as the number of molecules of product obtained when 100 could in principle have been formed. A yield of about 80 percent or more is generally considered good, but some transformations can prove so difficult to achieve that even a yield of 10 or 20 percent may have to be accepted. The ultimate synthetic goal in a perfect synthesis is to achieve 100 percent "atom efficiency," in which all atoms of all reagents are incorporated into the synthesized product without the formation of any by-products.

Naturally, the yield of a process affects the cost of the product, because the shortfall from a 100 percent yield represents wasted material. In addition, yield can be of the utmost importance in determining whether a synthesis is a practicable possibility, because the overall yield of a synthesis is the product of the yields of the individual steps.

If these intermediate yields are mostly low, the ultimate product may not be obtainable in the necessary amount from the available starting material.

Finally, consideration must be given to the rate at which each step in the planned sequence occurs. In many instances, a desired reaction is possible in principle but in practice takes place so slowly as to be ineffective. It is then necessary to investigate whether the rate can be increased to a practicable level by altering the conditions of the reaction—for example, by raising the temperature or by adding an extra species, called a catalyst, that increases the rate without altering the course of the reaction.

Isolation and Purification of Products

The product of a synthesis is normally contaminated with reagents used in the synthesis, by-products, and possibly some unchanged starting material; these contaminants must be removed in order for a pure product to be obtained. In a multistep synthesis, it is normally desirable to purify the product from each step before proceeding to the next.

References

- Chemical-engineering, technology: britannica.com, Retrieved 23 June, 2019
- What-is-a-chemical:thoughtco.com, Retrieved 12 April, 2019
- Chemical-element, science: britannica.com, Retrieved 21 March , 2019
- Chemical-compound, science: britannica.com, Retrieved 1 June, 2019

Branches of Chemical Engineering

2

- Polymer Engineering
- Chemical Reaction Engineering
- Chemical Thermodynamics
- Biochemical Engineering
- Biomolecular Engineering
- Pharmaceutical Engineering

There are various sub-disciplines of chemical engineering. These can be categorized into polymer engineering, chemical reaction engineering, biochemical engineering, pharmaceutical engineering, etc. This chapter closely examines these sub-disciplines of chemical engineering to provide an extensive understanding of the subject.

Polymer Engineering

Polymer engineering is an engineering that designs analysis and modify polymer material. Basic divisions of polymer are thermoplastic, elastomer, and thermoset which helps to design areas of application compounding and processing of polymer are major description of polymer. It includes different types of polymerization. The special technical importance of polymers can be judged by the fact that half of the professional organic chemists employed by industry in the United States are engaged in research or development related to polymers.

- Developing polymer processing
- Physical and organic chemistry
- Thermoplastics
- Fluid mechanics
- Polymerization

Chemical Reaction Engineering

Chemical reaction engineering (reaction engineering or reactor engineering) is a specialty in chemical engineering or industrial chemistry dealing with chemical reactors. Frequently the term relates specifically to catalytic reaction systems where either a homogeneous or heterogeneous catalyst is present in the reactor. Sometimes a reactor *per se* is not present by itself, but rather is integrated into a process, for example in reactive separations vessels, retorts, certain fuel cells, and photocatalytic surfaces. The issue of solvent effects on reaction kinetics is also considered as an integral part.

Chemical reaction engineering as a discipline started in the early 1950s under the impulse of researchers at the Shell Amsterdam research center and the university of Delft. The term chemical reaction engineering was apparently coined by J.C. Vlugter while preparing the 1st European Symposium on Chemical Reaction Engineering which was held in Amsterdam in 1957.

Chemical reaction engineering aims at studying and optimizing chemical reactions in order to define the best reactor design. Hence, the interactions of flow phenomena, mass transfer, heat transfer, and reaction kinetics are of prime importance in order to relate reactor performance to feed composition and operating conditions. Although originally applied to the petroleum and petrochemical industries, its general methodology combining reaction chemistry and chemical engineering concepts allows optimization of a variety of systems where modeling or engineering of reactions is needed. Chemical reaction engineering approaches are indeed tailored for the development of new processes and the improvement of existing technologies.

Chemical Thermodynamics

Chemical thermodynamics is the study of the interrelation of heat and work with chemical reactions or with physical changes of state within the confines of the laws of thermodynamics. Chemical thermodynamics involves not only laboratory measurements of various thermodynamic properties, but also the application of mathematical methods to the study of chemical questions and the *spontaneity* of processes.

The structure of chemical thermodynamics is based on the first two laws of thermodynamics. Starting from the first and second laws of thermodynamics, four equations called the "fundamental equations of Gibbs" can be derived. From these four, a multitude of equations, relating the thermodynamic properties of the thermodynamic system can be derived using relatively simple mathematics. This outlines the mathematical framework of chemical thermodynamics.

The primary objective of chemical thermodynamics is the establishment of a criterion for determination of the feasibility or spontaneity of a given transformation. In this manner, chemical thermodynamics is typically used to predict the energy exchanges that occur in the following processes:

- Chemical reactions.
- Phase changes.
- The formation of solutions.

The following state functions are of primary concern in chemical thermodynamics:

- Internal energy (U).
- Enthalpy (H).
- Entropy (S).
- Gibbs free energy (G).

Most identities in chemical thermodynamics arise from application of the first and second laws of thermodynamics, particularly the law of conservation of energy, to these state functions.

The 3 laws of thermodynamics:

- The energy of the universe is constant.
- In any spontaneous process, there is always an increase in entropy of the universe.
- The entropy of a perfect crystal (well ordered) at 0 Kelvin is zero.

Chemical Energy

Chemical energy is the potential of a chemical substance to undergo a transformation through a chemical reaction or to transform other chemical substances. Breaking or making of chemical bonds involves energy or heat, which may be either absorbed or evolved from a chemical system.

Energy that can be released (or absorbed) because of a reaction between a set of chemical substances is equal to the difference between the energy content of the products and the reactants. This change in energy is called the change in internal energy of a chemical reaction. Where $\Delta_f U^o_{reactants}$ is the internal energy of formation of the reactant molecules that can be calculated from the bond energies of the various chemical bonds of the molecules under consideration and $\Delta_f U^o_{products}$ is the internal energy of formation of the product molecules. The change in internal energy is a process which is equal to the heat change if it is measured under conditions of constant volume (at STP condition), as in a closed rigid container such as a bomb calorimeter. However, under conditions of constant pressure, as in reactions in vessels open to the atmosphere, the measured heat change is not always equal to the internal energy change, because pressure-volume work also releases or absorbs energy. (The heat change at constant pressure is called the enthalpy change; in this case the enthalpy of formation).

Another useful term is the heat of combustion, which is the energy released due to a combustion reaction and often applied in the study of fuels. Food is similar to hydrocarbon fuel and carbohydrate fuels, and when it is oxidized, its caloric content is similar (though not assessed in the same way as a hydrocarbon fuel).

In chemical thermodynamics the term used for the chemical potential energy is chemical potential, and for chemical transformation an equation most often used is the Gibbs-Duhem equation.

Chemical Reactions

In most cases of interest in chemical thermodynamics there are internal degrees of freedom and processes, such as chemical reactions and phase transitions, which always create entropy unless they are at equilibrium, or are maintained at a "running equilibrium" through "quasi-static" changes by being coupled to constraining devices, such as pistons or electrodes, to deliver and receive external work. Even for homogeneous "bulk" materials, the free energy functions depend on the composition, as do all the extensive thermodynamic potentials, including the internal energy. If the quantities $\{ N_i \}$, the number of chemical species, are omitted from the formulae, it is impossible to describe compositional changes.

Gibbs Function or Gibbs Energy

For a "bulk" (unstructured) system they are the last remaining extensive variables. For an unstructured, homogeneous "bulk" system, there are still various *extensive* compositional variables $\{ N_i \}$ that G depends on, which specify the composition, the amounts of each chemical substance, expressed as the numbers of molecules present or (dividing by Avogadro's number = 6.023×10^{23}), the numbers of moles:

$$G = G(T, P, \{N_i\}).$$

For the case where only PV work is possible:

$$dG = SdT + VdP + \sum_i \mu_i dN_i$$

in which μ_i is the chemical potential for the i-th component in the system:

$$\mu_i = \left(\frac{\partial G}{\partial N_i}\right)_{T,P,j\neq i,etc.}$$

The expression for dG is especially useful at constant T and P, conditions which are easy to achieve experimentally and which approximates the condition in living creatures:

$$(dG)_{T,P} = \sum_i \mu_i \, dN_i.$$

Chemical Affinity

While this formulation is mathematically defensible, it is not particularly transparent since one does not simply add or remove molecules from a system. There is always a *process* involved in changing the composition; e.g., a chemical reaction (or many), or movement of molecules from one phase (liquid) to another (gas or solid). We should find a notation which does not seem to imply that the amounts of the components (N_i) can be changed independently. All real processes obey conservation of mass, and in addition, conservation of the numbers of atoms of each kind. Whatever molecules are transferred to or from should be considered part of the "system".

Consequently, we introduce an explicit variable to represent the degree of advancement of a process, a progress variable ξ for the *extent of reaction,* and to the use of the partial derivative $\partial G/\partial\xi$ (in place of the widely used "ΔG", since the quantity at issue is not a finite change). The result is an understandable expression for the dependence of dG on chemical reactions (or other processes). If there is just one reaction:

$$(dG)_{T,P} = \left(\frac{\partial G}{\partial \xi}\right)_{T,P} d\xi$$

If we introduce the *stoichiometric coefficient* for the *i-th* component in the reaction:

$$\upsilon_i = \partial N_i / \partial \xi$$

which tells how many molecules of i are produced or consumed, we obtain an algebraic expression for the partial derivative:

$$\left(\frac{\partial G}{\partial \xi}\right)_{T,P} = \sum_i \mu_i \upsilon_i = -\mathbb{A}$$

where, we introduce a concise and historical name for this quantity, the "affinity", symbolized by A, as introduced by Théophile de Donder in 1923. The minus sign comes from the fact the affinity was defined to represent the rule that spontaneous changes will ensue only when the change in the Gibbs free energy of the process is negative, meaning that the chemical species have a positive affinity for each other. The differential for G takes on a simple form which displays its dependence on compositional change:

$$(dG)_{T,P} = -A d\xi.$$

If there are a number of chemical reactions going on simultaneously, as is usually the case:

$$(dG)_{T,P} = -\sum_k A_k \, d\xi_k.$$

a set of reaction coordinates $\{ \xi_j \}$, avoiding the notion that the amounts of the components (N_i) can be changed independently. The expressions above are equal to zero at thermodynamic equilibrium, while in the general case for real systems, they are negative because all chemical reactions proceeding at a finite rate produce entropy. This can be made even more explicit by introducing the reaction rates $d\xi_j/dt$. For each and every physically independent process.

$$A\dot{\xi} \le 0$$

This is a remarkable result since the chemical potentials are intensive system variables, depending only on the local molecular milieu. They cannot "know" whether the temperature and pressure (or any other system variables) are going to be held constant over time. It is a purely local criterion and must hold regardless of any such constraints. Of course, it could have been obtained by taking partial derivatives of any of the other fundamental state functions, but nonetheless is a general criterion for ($-T$ times) the entropy production from that spontaneous process; or at least any part of it that is not captured as external work.

We now relax the requirement of a homogeneous "bulk" system by letting the chemical potentials and the affinity apply to any locality in which a chemical reaction (or any other process) is occurring. By accounting for the entropy production due to irreversible processes, the inequality for dG is now replaced by an equality:

$$dG = -S\,dT + VdP - \sum_k A_k \, d\xi_k + W'$$

or

$$dG_{T,P} = -\sum_k A_k \, d\xi_k + W'.$$

Any decrease in the Gibbs function of a system is the upper limit for any isothermal, isobaric work that can be captured in the surroundings, or it may simply be dissipated, appearing as T times a corresponding increase in the entropy of the system and/or its surrounding. Or it may go partly toward doing external work and partly toward creating entropy. The important point is that the extent of reaction for a chemical reaction may be coupled to the displacement of some external mechanical or electrical quantity in such a way that one can advance only if the other one also does. The coupling may occasionally be rigid, but it is often flexible and variable.

Solutions

In solution chemistry and biochemistry, the Gibbs free energy decrease ($\partial G/\partial \xi$, in molar units, denoted cryptically by ΔG) is commonly used as a surrogate for ($-T$ times) the entropy produced by spontaneous chemical reactions in situations where there is no work being done; or at least no "useful" work; i.e., other than perhaps some $\pm P\,dV$. The assertion that all spontaneous reactions have a negative ΔG is merely a restatement of the fundamental thermodynamic relation, giving it the physical dimensions of energy and somewhat obscuring its significance in terms of entropy. When there is no useful work being done, it would be less misleading to use the Legendre transforms of the entropy appropriate for constant T, or for constant T and P, the Massieu functions $-F/T$ and $-G/T$ respectively.

Non Equilibrium

Generally the systems treated with the conventional chemical thermodynamics are either at equilibrium or near equilibrium. Ilya Prigogine developed the thermodynamic treatment of open systems that are far from equilibrium. In doing so he has discovered phenomena and structures of completely new and completely unexpected types. His generalized, nonlinear and irreversible thermodynamics has found surprising applications in a wide variety of fields.

The non equilibrium thermodynamics has been applied for explaining how ordered structures e.g. the biological systems, can develop from disorder. Even if Onsager's relations are utilized, the classical principles of equilibrium in thermodynamics still show that linear systems close to equilibrium always develop into states of disorder which are stable to perturbations and cannot explain the occurrence of ordered structures.

Prigogine called these systems dissipative systems, because they are formed and maintained by the dissipative processes which take place because of the exchange of energy between the system and its environment and because they disappear if that exchange ceases. They may be said to live in symbiosis with their environment.

The method which Prigogine used to study the stability of the dissipative structures to perturbations is of very great general interest. It makes it possible to study the most varied problems, such as city traffic problems, the stability of insect communities, the development of ordered biological structures and the growth of cancer cells to mention but a few examples.

System Constraints

In this regard, it is crucial to understand the role of walls and other *constraints*, and the distinction between *independent* processes and *coupling*. Contrary to the clear implications of many reference sources, the previous analysis is not restricted to homogeneous, isotropic bulk systems which can deliver only *PdV* work to the outside world, but applies even to the most structured systems. There are complex systems with many chemical "reactions" going on at the same time, some of which are really only parts of the same, overall process. An *independent* process is one that *could* proceed even if all others were unaccountably stopped in their tracks. Understanding this is perhaps a "thought experiment" in chemical kinetics, but actual examples exist.

A gas reaction which results in an increase in the number of molecules will lead to an increase in volume at constant external pressure. If it occurs inside a cylinder closed with a piston, the equilibrated reaction can proceed only by doing work against an external force on the piston. The extent variable for the reaction can increase only if the piston moves, and conversely, if the piston is pushed inward, the reaction is driven backwards.

Similarly, a redox reaction might occur in an electrochemical cell with the passage of current in wires connecting the electrodes. The half-cell reactions at the electrodes are constrained if no current is allowed to flow. The current might be dissipated as joule heating, or it might in turn run an electrical device like a motor doing mechanical work. An automobile lead-acid battery can be recharged, driving the chemical reaction backwards. In this case as well, the reaction is not an independent process. Some, perhaps most, of the Gibbs free energy of reaction may be delivered as external work.

The hydrolysis of ATP to ADP and phosphate can drive the force times distance work delivered by living muscles, and synthesis of ATP is in turn driven by a redox chain in mitochondria and chloroplasts, which involves the transport of ions across the membranes of these cellular organelles. The coupling of processes here, and in the previous examples, is often not complete. Gas can leak slowly past a piston, just as it can slowly leak out of a rubber balloon. Some reaction may occur in a battery even if no external current is flowing. There is usually a coupling coefficient, which may depend on relative rates, which determines what percentage of the driving free energy is turned into external work, or captured as "chemical work"; a misnomer for the free energy of another chemical process.

Biochemical Engineering

Biochemical engineering, also known as bioprocess engineering, is a field of study with roots stemming from chemical engineering and biological engineering. It mainly deals with the design, construction, and advancement of unit processes that involve biological organisms or organic molecules and has various applications in areas of interest such as biofuels, food, pharmaceuticals, biotechnology, and water treatment processes. The role of a biochemical engineer is to take findings developed by biologists and chemists in a laboratory and translate that to a large-scale manufacturing process.

Applications

Biotechnology

Biotechnology and biochemical engineering are closely related to each other as biochemical engineering can be considered a sub-branch of biotechnology. One of the primary focuses of biotechnology is in the medical field, where biochemical engineers work to design pharmaceuticals, artificial organs, biomedical devices, chemical sensors, and drug delivery systems. Biochemical engineers use their knowledge of chemical processes in biological systems in order to create tangible products that improve people's health. Specific areas of studies include metabolic, enzyme, and tissue engineering. The study of cell cultures is widely used in biochemical engineering and biotechnology due to its many applications in developing natural fuels, improving the efficiency in producing drugs and pharmaceutical processes, and also creating cures for disease. Other medical applications of biochemical engineering within biotechnology are genetics testing and pharmacogenomics.

Food Industry

Biochemical engineers primarily focus on designing systems that will improve the production, processing, packaging, storage, and distribution of food. Some commonly processed foods include wheat, fruits, and milk which undergo processes such as milling, dehydration, and pasteurization in order to become products that can be sold. There are three levels of food processing: primary, secondary, and tertiary. Primary food processing involves turning agricultural products into other products that can be turned into food, secondary food processing is the making of food from readily available ingredients, and tertiary food processing is commercial production of ready-to eat or heat-and-serve foods. Drying, pickling, salting, and fermenting foods were some of the oldest food processing techniques used to preserve food by preventing yeasts, molds, and bacteria to cause spoiling. Methods for preserving food have evolved to meet current standards of food safety but still use the same processes as the past. Biochemical engineers also work to improve the nutritional value of food products, such as in golden rice, which was developed to prevent vitamin A deficiency in certain areas where this was an issue. Efforts to advance preserving technologies can also ensure lasting retention of nutrients as foods are stored. Packaging plays a key role in preserving as well as ensuring the safety of the food by protecting the product from contamination, physical damage, and tampering. Packaging can also make it easier to transport and serve food. A common job for biochemical engineers working in the food industry is to design ways to perform all these processes on a large scale in order to meet the demands of the population. Responsibilities for this career path include designing and performing experiments, optimizing processes, consulting with groups to develop new technologies, and preparing project plans for equipment and facilities.

Biomolecular Engineering

Biomolecular engineering is the application of engineering principles and practices to the purposeful manipulation of molecules of biological origin. Biomolecular engineers integrate knowledge of biological processes with the core knowledge of chemical engineering in order to focus on molecular level solutions to issues and problems in the life sciences related to the environment, agriculture, energy, industry, food production, biotechnology and medicine.

Biomolecular engineers purposefully manipulate carbohydrates, proteins, nucleic acids and lipids within the framework of the relation between their structure, function and properties and in relation to applicability to such areas as environmental remediation, crop and livestock production, biofuel cells and biomolecular diagnostics. The thermodynamics and kinetics of molecular recognition in enzymes, antibodies, DNA hybridization, bio-conjugation/bio-immobilization and bioseparations are studied. Attention is also given to the rudiments of engineered biomolecules in cell signaling, cell growth kinetics, biochemical pathway engineering and bioreactor engineering.

Basic Biomolecules

Biomolecular engineering deals with the manipulation of many key biomolecules. These include, but are not limited to, proteins, carbohydrates, nucleic acids, and lipids. These molecules are the basic building blocks of life and by controlling, creating, and manipulating their form and function there are many new avenues and advantages available to society. Since every biomolecule is different, there are a number of techniques used to manipulate each one respectively.

Proteins

Proteins are polymers that are made up of amino acid chains linked with peptide bonds. They have four distinct levels of structure: primary, secondary, tertiary, and quaternary. Primary structure refers to the amino acid backbone sequence. Secondary structure focuses on minor conformations that develop as a result of the hydrogen bonding between the amino acid chain. If most of the protein contains intermolecular hydrogen bonds it is said to be fibrillar, and the majority of its secondary structure will be beta sheets. However, if the majority of the orientation contains intramolecular hydrogen bonds, then the protein is referred to as globular and mostly consists of alpha helices.

There are also conformations that consist of a mix of alpha helices and beta sheets as well as a beta helixes with an alpha sheets.

The tertiary structure of proteins deal with their folding process and how the overall molecule is arranged. Finally, a quaternary structure is a group of tertiary proteins coming together and binding. With all of these levels, proteins have a wide variety of places in which they can be manipulated and adjusted. Techniques are used to affect the amino acid sequence of the protein (site-directed mutagenesis), the folding and conformation of the protein, or the folding of a single tertiary protein within a quaternary protein matrix. Proteins that are the main focus of manipulation are typically enzymes. These are proteins that act as catalysts for biochemical reactions. By manipulating these catalysts, the reaction rates, products, and effects can be controlled. Enzymes and proteins are important to the biological field and research that there are specific divisions of engineering focusing only on proteins and enzymes.

Carbohydrates

Carbohydrates are another important biomolecule. These are polymers, called polysaccharides, which are made up of chains of simple sugars connected via glycosidic bonds. These monosaccharides consist of a five to six carbon ring that contains carbon, hydrogen, and oxygen - typically in a 1:2:1 ratio, respectively. Common monosaccharides are glucose, fructose, and ribose. When linked together monosaccharides can form disaccharides, oligosaccharides, and polysaccharides: the nomenclature is dependent on the number of monosaccharides linked together. Common dissacharides, two monosaccharides joined together, are sucrose, maltose, and lactose. Important polysaccharides, links of many monosaccharides, are cellulose, starch, and chitin.

Cellulose is a polysaccharide made up of beta 1-4 linkages between repeat glucose monomers. It is the most abundant source of sugar in nature and is a major part of the paper industry. Starch is also a polysaccharide made up of glucose monomers; however, they are connected via an alpha 1-4 linkage instead of beta. Starches, particularly amylase, are important in many industries, including the paper, cosmetic, and food. Chitin is a derivation of cellulose, possessing an acetamide group instead of an –OH on one of its carbons. Acetimide group is deacetylated the polymer chain is then called chitosan. Both of these cellulose derivatives are a major source of research for the biomedical and food industries. They have been shown to assist with blood clotting, have antimicrobial properties, and dietary applications. A lot of engineering and research is focusing on the degree of deacetylation that provides the most effective result for specific applications.

Nucleic Acids

Nucleic acids are macromolecules that consist of DNA and RNA which are biopolymers consisting of chains of biomolecules. These two molecules are the genetic code and template that make life possible. Manipulation of these molecules and structures causes major changes in function and expression of other macromolecules. Nucleosides are glycosylamines containing a nucleobase bound to either ribose or deoxyribose sugar via a beta-glycosidic linkage. The sequence of the bases determine the genetic code. Nucleotides are nucleosides that are phosphorylated by specific kinases via a phosphodiester bond. Nucleotides are the repeating structural units of nucleic acids. The nucleotides are made of a nitrogenous base, a pentose (ribose for RNA or deoxyribose for DNA), and three phosphate groups.

Lipids

Lipids are biomolecules that are made up of glycerol derivatives bonded with fatty acid chains. Glycerol is a simple polyol that has a formula of $C_3H_5(OH)_3$. Fatty acids are long carbon chains that have a carboxylic acid group at the end. The carbon chains can be either saturated with hydrogen; every carbon bond is occupied by a hydrogen atom or a single bond to another carbon in the carbon chain, or they can be unsaturated; namely, there are double bonds between the carbon atoms in the chain. Common fatty acids include lauric acid, stearic acid, and oleic acid. The study and engineering of lipids typically focuses on the manipulation of lipid membranes and encapsulation. Cellular membranes and other biological membranes typically consist of a phospholipid bilayer membrane, or a derivative thereof. Along with the study of cellular membranes, lipids are also important molecules for energy storage. By utilizing encapsulation properties and thermodynamic characteristics, lipids become significant assets in structure and energy control when engineering molecules.

Recombinant DNA

Recombinant DNA are DNA biomolecules that contain genetic sequences that are not native to the organism's genome. Using recombinant techniques, it is possible to insert, delete, or alter a DNA sequence precisely without depending on the location of restriction sites. Recombinant DNA is used for a wide range of applications.

Method

The traditional method for creating recombinant DNA typically involves the use of plasmids in the host bacteria. The plasmid contains a genetic sequence corresponding to the recognition site of a restriction endonuclease, such as EcoR1. After foreign DNA fragments, which have also been cut with the same restriction endonuclease, have been inserted into host cell, the restriction endonuclease gene is expressed by applying heat, or by introducing a biomolecule, such as arabinose. Upon expression, the enzyme will cleave the plasmid at its corresponding recognition site creating sticky ends on the plasmid. Ligases then joins the sticky ends to the corresponding sticky ends of the foreign DNA fragments creating a recombinant DNA plasmid.

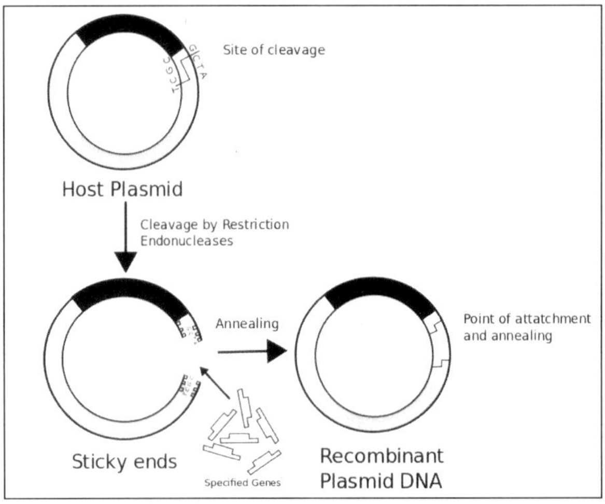

Creating recombinant DNA. After the plasmid is cleaved by restriction enzymes, ligases insert the foreign DNA fragments into the plasmid.

Advances in genetic engineering have made the modification of genes in microbes quite efficient allowing constructs to be made in about a weeks worth of time. It has also made it possible to modify the organism's genome itself. Specifically, use of the genes from the bacteriophage lambda are used in recombination. This mechanism, known as recombineering, utilizes the three proteins Exo, Beta, and Gam, which are created by the genes exo, bet, and gam respectively. Exo is a double stranded DNA exonuclease with 5' to 3' activity. It cuts the double stranded DNA leaving 3' overhangs. Beta is a protein that binds to single stranded DNA and assists homologous recombination by promoting annealing between the homology regions of the inserted DNA and the chromosomal DNA. Gam functions to protect the DNA insert from being destroyed by native nucleases within the cell.

Applications

Recombinant DNA can be engineered for a wide variety of purposes. The techniques utilized allow for specific modification of genes making it possible to modify any biomolecule. It can be engineered for laboratory purposes, where it can be used to analyze genes in a given organism. In the pharmaceutical industry, proteins can be modified using recombination techniques. Some of these proteins include human insulin. Recombinant insulin is synthesized by inserting the human insulin gene into *E. coli*, which then produces insulin for human use. Other proteins, such as human growth hormone, factor VIII, and hepatitis B vaccine are produced using similar means. Recombinant DNA can also be used for diagnostic methods involving the use of the ELISA method. This makes it possible to engineer antigens, as well as the enzymes attached, to recognize different substrates or be modified for bioimmobilization. Recombinant DNA is also responsible for many products found in the agricultural industry. Genetically modified food, such as golden rice, has been engineered to have increased production of vitamin A for use in societies and cultures where dietary vitamin A is scarce.

Other properties that have been engineered into crops include herbicide-resistance and insect-resistance.

Site-directed Mutagenesis

Site-directed mutagenesis is a technique that has been around since the 1970s. The early days of research in this field yielded discoveries about the potential of certain chemicals such as bisulfite and aminopurine to change certain bases in a gene. This research continued, and other processes were developed to create certain nucleotide sequences on a gene, such as the use of restriction enzymes to fragment certain viral strands and use them as primers for bacterial plasmids. The modern method, developed by Michael Smith in 1978, uses an oligonucleotide that is complementary to a bacterial plasmid with a single base pair mismatch or a series of mismatches.

Procedure

Site directed mutagenesis is a valuable technique that allows for the replacement of a single base in an oligonucleotide or gene. The basics of this technique involve the preparation of a primer that will be a complementary strand to a wild type bacterial plasmid. This primer will have a base pair mismatch at the site where the replacement is desired. The primer must also be long enough such that the primer will anneal to the wild type plasmid. After the primer anneals, a DNA polymerase will complete the primer. When the bacterial plasmid is replicated, the mutated strand will be replicated as well. The same technique can be used to create a gene insertion or deletion. Often, an antibiotic resistant gene is inserted along with the modification of interest and the bacteria are cultured on an antibiotic medium. The bacteria that were not successfully mutated will not survive on this medium, and the mutated bacteria can easily be cultured.

Applications

Site-directed mutagenesis can be useful for many different reasons. A single base pair replacement, could change a codon, and thus replace an amino acid in a protein. This is useful for studying the way certain proteins behave. It is also useful because enzymes can be purposefully manipulated by changing certain amino acids. If an amino acid is changed that is in close proximity to the active site, the kinetic parameters may change drastically, or the enzyme might behave in a different way. Another application of site directed mutagenesis is exchanging an amino acid residue far from the active site with a lysine residue or cysteine residue. These amino acids make it easier to covalently bond the enzyme to a solid surface, which allows for enzyme re-use and use of enzymes in continuous processes. Sometimes, amino acids with non-natural functional groups (such as ketones and azides) are added to proteins These additions may be for ease of bioconjugation, or to study the effects of amino acid changes on the form and function of the proteins. The coupling of site directed mutagenesis and PCR are being utilized to reduce interleukin-6 activity in cancerous cells. The bacteria *bacillus subtilis* is often used in site directed mutagenesis. The bacteria secretes an enzyme called subtilisin through the cell wall. Biomolecular engineers can purposely manipulate this gene to essentially make the cell a factory for producing whatever protein the insertion in the gene codes.

Bio-immobilization and Bio-conjugation

Bio-immobilization and bio-conjugation is the purposeful manipulation of a biomolecule's mobility by chemical or physical means to obtain a desired property. Immobilization of biomolecules allows exploiting characteristics of the molecule under controlled environments. For example , the immobilization of glucose oxidase on calcium alginate gel beads can be used in a bioreactor. The resulting product will not need purification to remove the enzyme because it will remain linked to the beads in the column. Examples of types of biomolecules that are immobilized are enzymes, organelles, and complete cells. Biomolecules can be immobilized using a range of techniques. The most popular are physical entrapment, adsorption, and covalent modification.

- Physical entrapment - the use of a polymer to contain the biomolecule in a matrix without chemical modification. Entrapment can be between lattices of polymer, known as gel entrapment, or within micro-cavities of synthetic fibers, known as fiber entrapment. Examples include entrapment of enzymes such as glucose oxidase in gel column for use as a bioreactor. Important characteristic with entrapment is biocatalyst remains structurally unchanged, but creates large diffusion barriers for substrates.

- Adsorption- immobilization of biomolecules due to interaction between the biomolecule and groups on support. Can be physical adsorption, ionic bonding, or metal binding chelation. Such techniques can be performed under mild conditions and relatively simple, although the linkages are highly dependent upon pH, solvent and temperature. Examples include enzyme-linked immunosorbent assays.

- Covalent modification- involves chemical reactions between certain functional groups and matrix. This method forms stable complex between biomolecule and matrix and is suited for mass production. Due to the formation of chemical bond to functional groups, loss of activity can occur. Examples of chemistries used are DCC coupling PDC coupling and EDC/NHS coupling, all of which take advantage of the reactive amines on the biomolecule's surface.

Because immobilization restricts the biomolecule, care must be given to ensure that functionality is not entirely lost. Variables to consider are pH, temperature, solvent choice, ionic strength, orientation of active sites due to conjugation. For enzymes, the conjugation will lower the kinetic rate due to a change in the 3-dimensional structure, so care must be taken to ensure functionality is not lost. Bio-immobilization is used in technologies such as diagnostic bioassays, biosensors, ELISA, and bioseparations.

Interleukin (IL-6) can also be bioimmobilized on biosensors. The ability to observe these changes in IL-6 levels is important in diagnosing an illness. A cancer patient will have elevated IL-6 level and monitoring those levels will allow the physician to watch the disease progress. A direct immobilization of IL-6 on the surface of a biosensor offers a fast alternative to ELISA.

Polymerase Chain Reaction

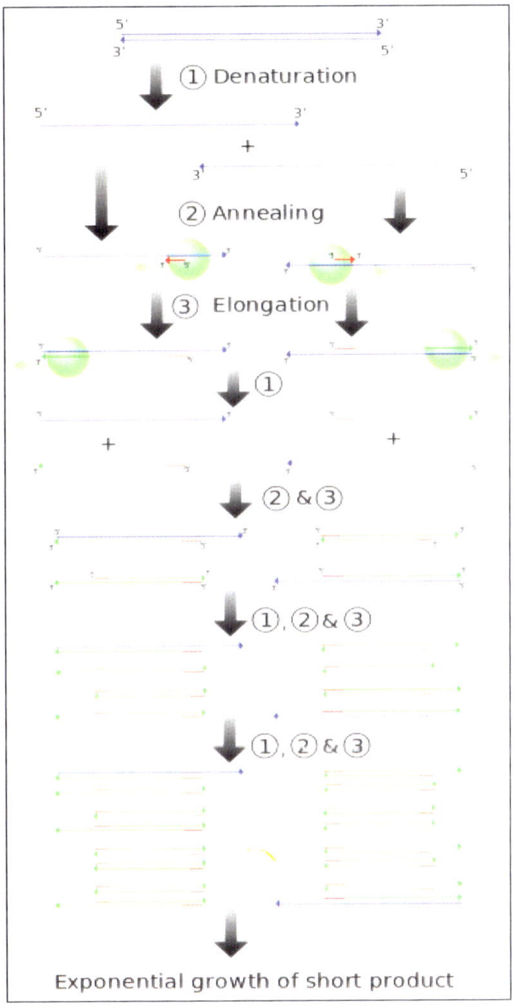

There are three main steps involved in PCR. In the first step, the double stranded DNA strands are "melted" or denatured forming single stranded DNA. Next, primers, which have been designed to target a specific gene sequence on the DNA, anneal to the single stranded DNA. Lastly, DNA polymerase synthesizes a new DNA strand complementary to the original DNA. These three steps are repeated multiple times until the desired number of copies are made.

The polymerase chain reaction (PCR) is a scientific technique that is used to replicate a piece of a DNA molecule by several orders of magnitude. PCR implements a cycle of repeated heated and cooling known as thermal cycling along with the addition of DNA primers and DNA polymerases to selectively replicate the DNA fragment of interest. The technique was developed by Kary Mullis in 1983 while working for the Cetus Corporation. Mullis would go on to win the Nobel Prize in Chemistry in 1993 as a result of the impact that PCR had in many areas such as DNA cloning, DNA sequencing, and gene analysis.

Biomolecular Engineering Techniques Involved in PCR

A number of biomolecular engineering strategies have played a very important role in the development and practice of PCR. For instance a crucial step in ensuring the accurate replication of the desired DNA fragment is the creation of the correct DNA primer. The most common method of primer synthesis is by the phosphoramidite method. This method includes the biomolecular engineering of a number of molecules to attain the desired primer sequence. The most prominent biomolecular engineering technique seen in this primer design method is the initial bioimmobilization of a nucleotide to a solid support. This step is commonly done via the formation of a covalent bond between the 3'-hydroxy group of the first nucleotide of the primer and the solid support material.

Furthermore, as the DNA primer is created certain functional groups of nucleotides to be added to the growing primer require blocking to prevent undesired side reactions. This blocking of functional groups as well as the subsequent de-blocking of the groups, coupling of subsequent nucleotides, and eventual cleaving from the solid support are all methods of manipulation of biomolecules that can be attributed to biomolecular engineering. The increase in interleukin levels is directly proportional to the increased death rate in breast cancer patients. PCR paired with Western blotting and ELISA help define the relationship between cancer cells and IL-6.

Enzyme-linked Immunosorbent Assay (ELISA)

Enzyme-linked immunosorbent assay is an assay that utilizes the principle of antibody-antigen recognition to test for the presence of certain substances. The three main types of ELISA tests which are indirect ELISA, sandwich ELISA, and competitive ELISA all rely on the fact that antibodies have an affinity for only one specific antigen. Furthermore, these antigens or antibodies can be attached to enzymes which can react to create a colorimetric result indicating the presence of the antibody or antigen of interest. Enzyme linked immunosorbent assays are used most commonly as diagnostic tests to detect HIV antibodies in blood samples to test for HIV, human chorionic gonadotropin molecules in urine to indicate pregnancy, and Mycobacterium tuberculosis antibodies in blood to test patients for tuberculosis. Furthermore, ELISA is also widely used as a toxicology screen to test people's serum for the presence of illegal drugs.

Techniques Involved in ELISA

Although there are three different types of solid state enzyme-linked immunosorbent assays, all three types begin with the bioimmobilization of either an antibody or antigen to a surface. This bioimmobilization is the first instance of biomolecular engineering that can be seen in ELISA implementation. This step can be performed in a number of ways including a covalent linkage to a surface which may be coated with protein or another substance. The bioimmobilization can also be performed via hydrophobic interactions between the molecule and the surface. Because there are many different types of ELISAs used for many different purposes the biomolecular engineering that this step requires varies depending on the specific purpose of the ELISA.

Another biomolecular engineering technique that is used in ELISA development is the bioconjugation of an enzyme to either an antibody or antigen depending on the type of ELISA. There is much to consider in this enzyme bioconjugation such as avoiding interference with the active site of the enzyme as well as the antibody binding site in the case that the antibody is conjugated with enzyme. This bioconjugation is commonly performed by creating crosslinks between the two molecules of interest and can require a wide variety of different reagents depending on the nature of the specific molecules.

Interleukin (IL-6) is a signaling protein that has been known to be present during an immune response. The use of the sandwich type ELISA quantifies the presence of this cytokine within spinal fluid or bone marrow samples.

Applications and Fields

In Industry

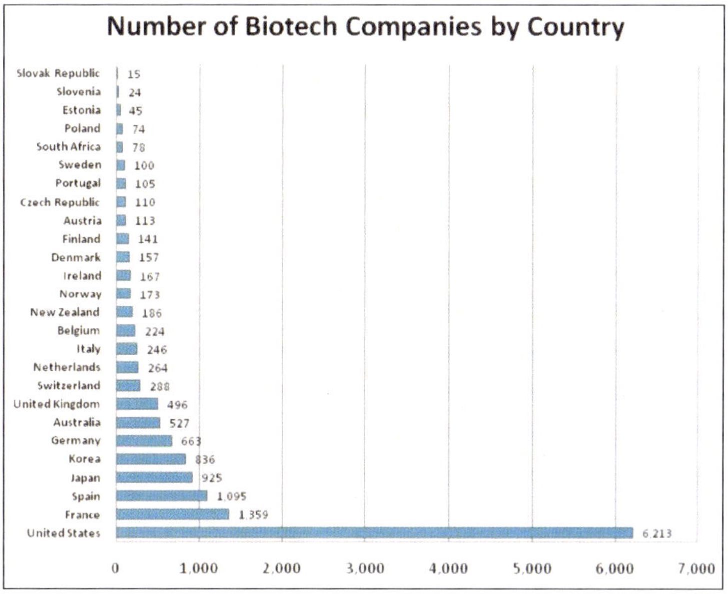

Graph showing number of biotech companies per country.

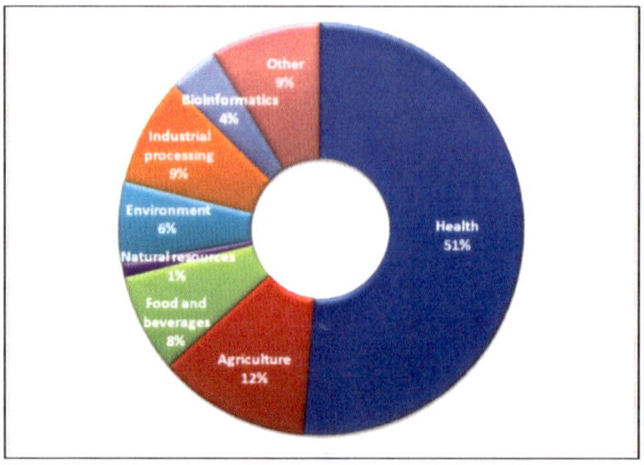

Graph showing percentages of biotech firms by application.

Biomolecular engineering is an extensive discipline with applications in many different industries and fields. As such, it is difficult to pinpoint a general perspective on the Biomolecular engineering profession. The biotechnology industry, however, provides an adequate representation. The biotechnology industry, or biotech industry, encompasses all firms that use biotechnology to produce goods or services or to perform biotechnology research and development. In this way, it encompasses many of the industrial applications of the biomolecular engineering discipline. By examination of the biotech industry, it can be gathered that the principal leader of the industry is the United States, followed by France and Spain. It is also true that the focus of the biotechnology industry and the application of biomolecular engineering is primarily clinical and medical. People are willing to pay for good health, so most of the money directed towards the biotech industry stays in health-related ventures.

Scale-up

Scaling up a process involves using data from an experimental-scale operation (model or pilot plant) for the design of a large (scaled-up) unit, of commercial size. Scaling up is a crucial part of commercializing a process. For example, insulin produced by genetically modified Escherichia coli bacteria was initialized on a lab-scale, but to be made commercially viable had to be scaled up to an industrial level. In order to achieve this scale-up a lot of lab data had to be used to design commercial sized units. For example, one of the steps in insulin production involves the crystallization of high purity glargin insulin. In order to achieve this process on a large scale we want to keep the Power/Volume ratio of both the lab-scale and large-scale crystallizers the same in order to achieve homogeneous mixing. We also assume the lab-scale crystallizer has geometric similarity to the large-scale crystallizer. Therefore,

$$P/V \ \alpha \ N_i^3 d_i^3$$

where,

- d_i= crystallizer impeller diameter.
- N_i= impeller rotation rate.

Pharmaceutical Engineering

Pharmaceutical engineering is a branch of pharmaceutical science and technology that involves development and manufacturing of products, processes, and components in the pharmaceuticals industry. While developing pharmaceutical products involves many interrelated disciplines like medicinal chemists, analytical chemists, clinicians/pharmacologists, pharmacists, chemical engineers, biomedical engineers, etc, the specific subfield of "pharmaceutical engineering" has only emerged recently as a distinct engineering discipline. This now brings the problem-solving principles and quantitative training of engineering to complement the other scientific fields already involved in drug development.

- Clinical and Biomedical Engineering.
- Cell and Tissue Engineering.
- Biomaterials and Biomechanics.
- Bioinformatics and Computational Biology.
- Genetic Engineering and Genomics.
- Neural and Rehabilitation Engineering.
- Bioinstrumentation.
- Medical Devices and Medical Imaging.
- Bio-pharmaceutics.

References

- Chemical-and-polymer-engineering, conferences-list: omicsonline.org, Retrieved 24 March , 2019
- Ott, bevan j.; boerio-goates, juliana (2000). Chemical thermodynamics – principles and applications. Academic press. Isbn 0-12-530990-2.
- R, kelly; health, jh bloomberg school of public. "food processing". Johns hopkins bloomberg school of public health. Retrieved 2019-03-18.
- Ryu, d. D. Y.; nam, d.-h. (2000). "recent progress in biomolecular engineering". Biotechnol progress. 16 (1): 2–16. Doi:10.1021/bp088059d. Pmid 10662483.
- Slabaugh, michael r. & seager, spencer l. (2007). Organic and biochemistry for today (6th ed.). Pacific grove: brooks cole. Isbn 978-0-495-11280-8.
- Pharmaceutical-engineering, events-list: pharmaceuticalconferences.com, Retrieved 13 July, 2019

Chemical Kinetics 3

- **Equilibrium Chemistry**
- **Chemical Reaction**
- **Law of Mass Action**
- **Chemical Equilibrium**
- **Collision Theory**

Chemical kinetics is defined as the branch of chemistry which deals with the understanding of the rates of chemical reactions. A few of its significant areas are collision theory, chemical equilibrium, law of mass action, etc. This chapter discusses in detail all these aspects related to chemical kinetics.

Chemical Kinetics

Chemical kinetics is the branch of physical chemistry that is concerned with understanding the rates of chemical reactions. It is to be contrasted with thermodynamics, which deals with the direction in which a process occurs but in itself tells nothing about its rate. Thermodynamics is time's arrow, while chemical kinetics is time's clock. Chemical kinetics relates to many aspects of cosmology, geology, biology, engineering, and even psychology and thus has far-reaching implications. The principles of chemical kinetics apply to purely physical processes as well as to chemical reactions.

One reason for the importance of kinetics is that it provides evidence for the mechanisms of chemical processes. Besides being of intrinsic scientific interest, knowledge of reaction mechanisms is of practical use in deciding what is the most effective way of causing a reaction to occur. Many commercial processes can take place by alternative reaction paths, and knowledge of the mechanisms makes it possible to choose reaction conditions that favour one path over others.

A chemical reaction is, by definition, one in which chemical substances are transformed into other substances, which means that chemical bonds are broken and formed so that there are changes in the relative positions of atoms in molecules. At the same time, there are shifts in the arrangements of the electrons that form the chemical bonds. A description of a reaction mechanism must therefore deal with the movements and speeds of atoms and electrons. The detailed mechanism by which a chemical process occurs is referred to as the reaction path, or pathway.

The vast amount of work done in chemical kinetics has led to the conclusion that some chemical reactions go in a single step; these are known as elementary reactions. Other reactions go in more than one step and are said to be stepwise, composite, or complex. Measurements of the rates of chemical reactions over a range of conditions can show whether a reaction proceeds by one or more steps. If a reaction is stepwise, kinetic measurements provide evidence for the mechanism of the individual elementary steps. Information about reaction mechanisms is also provided by certain nonkinetic studies, but little can be known about a mechanism until its kinetics has been investigated. Even then, some doubt must always remain about a reaction mechanism. An investigation, kinetic or otherwise, can disprove a mechanism but can never establish it with absolute certainty.

Equilibrium Chemistry

Equilibrium chemistry is concerned with systems in chemical equilibrium. The unifying principle is that the free energy of a system at equilibrium is the minimum possible, so that the slope of the free energy with respect to the reaction coordinate is zero. This principle, applied to mixtures at equilibrium provides a definition of an equilibrium constant. Applications include acid–base, host–guest, metal–complex, solubility, partition, chromatography and redox equilibria.

Thermodynamic Equilibrium

A chemical system is said to be in equilibrium when the quantities of the chemical entities involved do not and *cannot* change in time without the application of an external influence. In this sense a system in chemical equilibrium is in a stable state. The system at chemical equilibrium will be at a constant temperature, pressure (or volume) and composition. It will be insulated from exchange of heat with the surroundings, that is, it is a closed system. A change of temperature, pressure (or volume) constitutes an external influence and the equilibrium quantities will change as a result of such a change. If there is a possibility that the composition might change, but the rate of change is negligibly slow, the system is said to be in a metastable state. The equation of chemical equilibrium can be expressed symbolically as:

$$\text{reactant(s)} \rightleftharpoons \text{product(s)}$$

The sign $\rightleftharpoons$ means "are in equilibrium with". This definition refers to macroscopic properties. Changes do occur at the microscopic level of atoms and molecules, but to such a minute extent that they are not measurable and in a balanced way so that the macroscopic quantities do not change. Chemical equilibrium is a dynamic state in which forward and backward reactions proceed at such rates that the macroscopic composition of the mixture is constant. Thus, equilibrium sign $\rightleftharpoons$ symbolizes the fact that reactions occur in both forward $\rightharpoonup$ and backward $\leftharpoondown$ directions.

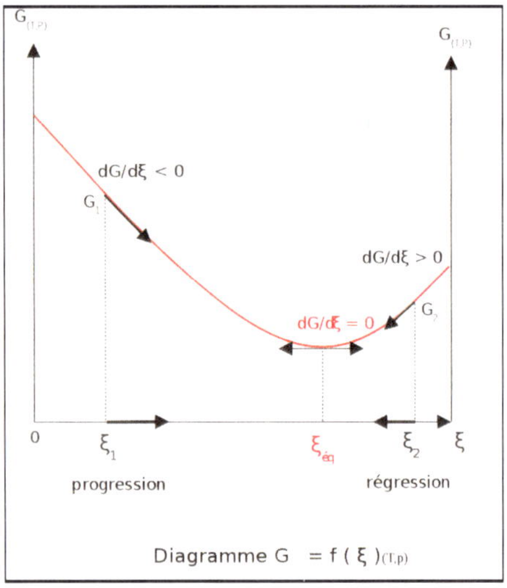

Diagramme G = f (ξ)(T,p)

A steady state, on the other hand, is not necessarily an equilibrium state in the chemical sense. For example, in a radioactive decay chain the concentrations of intermediate isotopes are constant because the rate of production is equal to the rate of decay. It is not a chemical equilibrium because the decay process occurs in one direction only.

Thermodynamic equilibrium is characterized by the free energy for the whole (closed) system being a minimum. For systems at constant volume the Helmholtz free energy is minimum and for systems at constant pressure the Gibbs free energy is minimum. Thus a metastable state is one for which the free energy change between reactants and products is not minimal even though the composition does not change in time.

The existence of this minimum is due to the free energy of mixing of reactants and products being always negative. For ideal solutions the enthalpy of mixing is zero, so the minimum exists because the entropy of mixing is always positive. The slope of the reaction free energy, δG_r with respect to the reaction coordinate, ξ, is zero when the free energy is at its minimum value.

$$\delta G_r = \left(\frac{\partial G}{\partial \xi}\right)_{T,P} \; ; \; \delta G_r(Eq) = 0$$

Equilibrium Constant

Chemical potential is the partial molar free energy. The potential, μ_i, of the ith species in a chemical reaction is the partial derivative of the free energy with respect to the number of moles of that species, N_i:

$$\mu_i = \left(\frac{\partial G}{\partial N_i}\right)_{T,P}$$

A general chemical equilibrium can be written as:

$$\sum_j n_j \,\text{Reactant}_j \rightleftharpoons \sum_k m_k \,\text{product}_k$$

n_j are the stoichiometric coefficients of the reactants in the equilibrium equation, and m_j are the coefficients of the products. The value of δG_r for these reactions is a function of the chemical potentials of all the species.

$$\delta G_r = \sum_k m_k \mu_k - \sum_j n_j \mu_j$$

The chemical potential, μ_i, of the *i*th species can be calculated in terms of its activity, a_i.

$$\mu_i = \mu_i^{\ominus} + RT \ln a_i$$

μ^{o}_i is the standard chemical potential of the species, *R* is the gas constant and *T* is the temperature. Setting the sum for the reactants *j* to be equal to the sum for the products, *k*, so that $\delta G_r(\text{Eq}) = 0$:

$$\sum_j n_j \left(\mu_j^{\ominus} + RT \ln a_j\right) = \sum_k m_k (\mu_k^{\ominus} + RT \ln a_k)$$

Rcarranging thc tcrms,

$$\sum_k m_k \mu_k^{\ominus} - \sum_j n_j \mu_j^{\ominus} = -RT\left(\sum_k \ln a_k^{m_k} - \sum_j \ln a_j^{n_j}\right)$$

$$\Delta G^{\ominus} = -RT \ln K.$$

This relates the standard Gibbs free energy change, ΔG^{o} to an equilibrium constant, *K*, the reaction quotient of activity values at equilibrium.

$$\Delta G^{\ominus} = \sum_k m_k \mu_k^{\ominus} - \sum_j n_j \mu_j^{\ominus}$$

$$\ln K = \sum_k \ln a_k^{m_k} - \sum_j \ln a_j^{n_j}; \quad K = \frac{\prod_k a_k^{m_k}}{\prod_j a_j^{n_j}}$$

It follows that any equilibrium of this kind can be characterized either by the standard free energy change or by the equilibrium constant. In practice concentrations are more useful than activities. Activities can be calculated from concentrations if the activity coefficient are known, but this is rarely the case. Sometimes activity coefficients can be calculated using, for example, Pitzer equations or Specific ion interaction theory. Otherwise conditions must be adjusted so that activity coefficients do not vary much. For ionic solutions this is achieved by using a background ionic medium at a high concentration relative to the concentrations of the species in equilibrium.

If activity coefficients are unknown they may be subsumed into the equilibrium constant, which becomes a concentration quotient. Each activity a_i is assumed to be the product of a concentration, $[A_i]$, and an activity coefficient, γ_i:

$$\tilde{a}_i = [\mathrm{A}_i]\gamma_i$$

This expression for activity is placed in the expression defining the equilibrium constant.

$$K = \frac{\prod_k \tilde{a}_k^{\,m_k}}{\prod_j a_j^{\,n_j}} = \frac{\prod_k ([\mathrm{A}_k]\ _k)^{m_k}}{\prod_j \tilde{a}_j^{\,n_j}} = \frac{\prod_k [\mathrm{A}_k]^{m_k}}{\prod_j [\mathrm{A}_j]^{n_j}} \times \frac{\prod_k \ ^{m_k}}{\prod_j \ _j^{\,n_j}} = \frac{\prod_k [\mathrm{A}_k]^{m_k}}{\prod_j [\mathrm{A}_j]^{n_j}} \times \Gamma$$

By setting the quotient of activity coefficients, Γ, equal to one, the equilibrium constant is defined as a quotient of concentrations.

$$K = \frac{\prod_k [\mathrm{A}_k]^{m_k}}{\prod_j [\mathrm{A}_j]^{n_j}}$$

In more familiar notation, for a general equilibrium:

$$\alpha\,\mathrm{A} + \beta\,\mathrm{B} \ldots \rightleftharpoons \sigma\,\mathrm{S} + \tau\,\mathrm{T} \ldots$$

$$K = \frac{[\mathrm{S}]^{\sigma}[\mathrm{T}]^{\tau}\cdots}{[\mathrm{A}]^{\alpha}[\mathrm{B}]^{\hat{a}}\cdots}$$

This definition is much more practical, but an equilibrium constant defined in terms of concentrations is dependent on conditions. In particular, equilibrium constants for species in aqueous solution are dependent on ionic strength, as the quotient of activity coefficients varies with the ionic strength of the solution.

The values of the standard free energy change and of the equilibrium constant are temperature dependent. To a first approximation, the van't Hoff equation may be used.

$$\frac{d\ln K}{dT} = \frac{\Delta H^{\ominus}}{RT^2} \quad \text{or} \quad \frac{d\ln K}{d\frac{1}{T}} = -\frac{\Delta H^{\ominus}}{R}$$

This shows that when the reaction is exothermic (ΔH°, the standard enthalpy change, is negative), then K decreases with increasing temperature, in accordance with Le Châtelier's principle. The approximation involved is that the standard enthalpy change, ΔH°, is independent of temperature, which is a good approximation only over a small temperature range. Thermodynamic arguments can be used to show that:

$$\left(\frac{\partial H}{\partial T}\right)_p = C_p$$

where C_p is the heat capacity at constant pressure.

Equilibria Involving Gases

When dealing with gases, fugacity, f, is used rather than activity. However, whereas activity is dimensionless, fugacity has the dimension of pressure. A consequence is that chemical potential has to be defined in terms of a standard pressure, p°:

$$\mu = \mu^{\ominus} + RT \ln \frac{f}{p^{\ominus}}$$

By convention p° is usually taken to be 1 bar. Fugacity can be expressed as the product of partial pressure, p, and a fugacity coefficient, Φ:

$$f = p\Phi$$

Fugacity coefficients are dimensionless and can be obtained experimentally at specific temperature and pressure, from measurements of deviations from ideal gas behaviour. Equilibrium constants are defined in terms of fugacity. If the gases are at sufficiently low pressure that they behave as ideal gases, the equilibrium constant can be defined as a quotient of partial pressures.

An example of gas-phase equilibrium is provided by the Haber–Bosch process of ammonia synthesis.

$$N_2 + 3\,H_2 \rightleftharpoons 2\,NH_3; \quad K = \frac{f\,NH_3^{\,2}}{f\,N_2\, f\,H_2^{\,3}}$$

This reaction is strongly exothermic, so the equilibrium constant decreases with temperature. However, a temperature of around 400 °C is required in order to achieve a reasonable rate of reaction with currently available catalysts. Formation of ammonia is also favoured by high pressure, as the volume decreases when the reaction takes place. The same reaction, nitrogen fixation, occurs at ambient temperatures in nature, when the catalyst is an enzyme such as nitrogenase. Much energy is needed initially to break the nitrogen–nitrogen triple bond even though the overall reaction is exothermic.

Gas-phase equilibria occur during combustion and were studied as early as 1943 in connection with the development of the V2 rocket engine.

The calculation of composition for a gaseous equilibrium at constant pressure is often carried out using ΔG values, rather than equilibrium constants.

Multiple Equilibria

Two or more equilibria can exist at the same time. When this is so, equilibrium constants can be ascribed to individual equilibria, but they are not always unique. For example, three equilibrium constants can be defined for a dibasic acid, H_2A.

$$A^{2-} + H^+ \rightleftharpoons HA^-;\ K_1 = \frac{[HA^-]}{[H^+][A^{2-}]}$$

$$HA^- + H^+ \rightleftharpoons H_2A;\ K_2 = \frac{[H_2A]}{[H^+]^2[A^-]}$$

$$A^{2-} + 2\,H^+ \rightleftharpoons H_2A;\ \beta_2 = \frac{[H_2A]}{[H^+]^2[A^{2-}]}$$

The three constants are not independent of each other and it is easy to see that $\beta_2 = K_1K_2$. The constants K_1 and K_2 are stepwise constants and β is an example of an overall constant.

Speciation

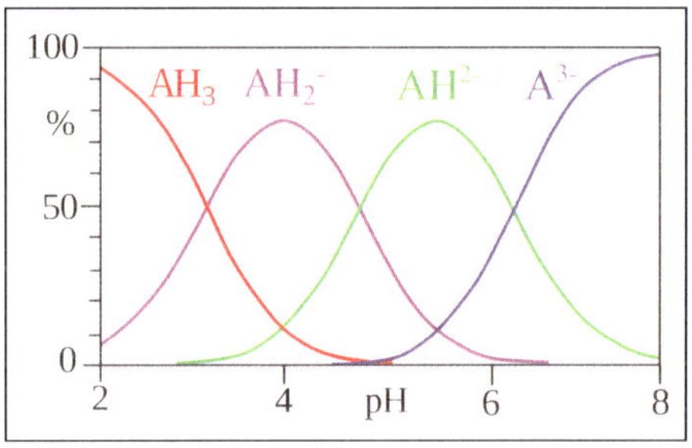

Speciation diagram for a solution of citric acid as a function of pH.

The concentrations of species in equilibrium are usually calculated under the assumption that activity coefficients are either known or can be ignored. In this case, each equilibrium constant for the formation of a complex in a set of multiple equilibria can be defined as follows:

$$\alpha\,A + \beta\,B \ldots \rightleftharpoons A_\alpha B_\beta \ldots;\ K_{\alpha\beta\ldots} = \frac{[A_\alpha B_\beta \ldots]}{[A]^\alpha [B]^\beta \ldots}$$

The concentrations of species containing reagent A are constrained by a condition of mass-balance, that is, the total (or analytical) concentration, which is the sum of all species' concentrations, must be constant. There is one mass-balance equation for each reagent of the type:

$$T_A = [A] + \sum [A_\alpha B_\beta \ldots] = [A] + \sum (\alpha K_{\alpha\beta\ldots} [A]^\alpha [B]^\beta \ldots)$$

There are as many mass-balance equations as there are reagents, A, B..., so if the equilibrium constant values are known, there are n mass-balance equations in n unknowns, [A], [B]..., the so-called free reagent concentrations. Solution of these equations gives all the information needed to calculate the concentrations of all the species.

Thus, the importance of an equilibrium constants lies in the fact that, once their values have been determined by experiment, they can be used to calculate the concentrations, known as the speciation, of mixtures that contain the relevant species.

Determination

There are five main types of experimental data that are used for the determination of solution equilibrium constants. Potentiometric data obtained with a glass electrode are the most widely used with aqueous solutions. The others are Spectrophotometric, Fluorescence (luminescence) measurements and NMR chemical shift measurements; simultaneous measurement of K and ΔH for 1:1 adducts in biological systems is routinely carried out using Isothermal Titration Calorimetry.

The experimental data will comprise a set of data points. At the i'th data point, the analytical concentrations of the reactants, $T_{A(i)}$, $T_{B(i)}$ etc. will be experimentally known quantities and there will be one or more measured quantities, y_i, that depend in some way on the analytical concentrations and equilibrium constants. A general computational procedure has three main components.

1. Definition of a chemical model of the equilibria. The model consists of a list of reagents, A, B, etc. and the complexes formed from them, with stoichiometries A_pB_q... Known or estimated values of the equilibrium constants for the formation of all complexes must be supplied.

2. Calculation of the concentrations of all the chemical species in each solution. The free concentrations are calculated by solving the equations of mass-balance, and the concentrations of the complexes are calculated using the equilibrium constant definitions. A quantity corresponding to the observed quantity can then be calculated using physical principles such as the Nernst potential or Beer-Lambert law which relate the calculated quantity to the concentrations of the species.

3. Refinement of the equilibrium constants. Usually a Non-linear least squares procedure is used. A weighted sum of squares, U, is minimized.

$$U = \sum_{i=np}^{i=1} w_i \left(y_i^{\text{observed}} - y_i^{\text{calculated}} \right)^2$$

The weights, w_i and quantities y may be vectors. Values of the equilibrium constants are refined in an iterative procedure.

Acid–base Equilibria

Brønsted and Lowry characterized an acid–base equilibrium as involving a proton exchange reaction:

acid + base $\rightleftharpoons$ conjugate base + conjugate acid.

An acid is a proton donor; the proton is transferred to the base, a proton acceptor, creating a conjugate acid. For aqueous solutions of an acid HA, the base is water; the conjugate base is A^- and the conjugate acid is the solvated hydrogen ion. In solution chemistry, it is usual to use H^+ as an abbreviation for the solvated hydrogen ion, regardless of the solvent. In aqueous solution H^+ denotes a solvated hydronium ion.

The Brønsted–Lowry definition applies to other solvents, such as dimethyl sulfoxide: the solvent S acts as a base, accepting a proton and forming the conjugate acid SH^+. A broader definition of acid dissociation includes hydrolysis, in which protons are produced by the splitting of water molecules. For example, boric acid, $B(OH)_3$, acts as a weak acid, even though it is not a proton donor, because of the hydrolysis equilibrium:

$$B(OH)_3 + H_2O \rightleftharpoons B(OH)^-_4 + H^+.$$

Similarly, metal ion hydrolysis causes ions such as $[Al(H_2O)_6]^{3+}$ to behave as weak acids:

$$[Al(H_2O)_6]^{3+} \rightleftharpoons [Al(H_2O)_5(OH)]^{2+} + H^+.$$

Acid–base equilibria are important in a very wide range of applications, such as acid–base homeostasis, ocean acidification, pharmacology and analytical chemistry.

Host–guest Equilibria

A host–guest complex, also known as a donor–acceptor complex, may be formed from a Lewis base, B, and a Lewis acid, A. The host may be either a donor or an acceptor. In biochemistry host–guest complexes are known as receptor-ligand complexes; they are formed primarily by non-covalent bonding. Many host–guest complexes has 1:1 stoichiometry, but many others have more complex structures. The general equilibrium can be written as:

$$p\,A + q\,B \rightleftharpoons A_pB_q$$

The study of these complexes is important for supramolecular chemistry and molecular recognition. The objective of these studies is often to find systems with a high binding selectivity of a host (receptor) for a particular target molecule or ion, the guest or ligand. An application is the development of chemical sensors. Finding a drug which either blocks a receptor, an antagonist which forms a strong complex the receptor, or activate it, an agonist, is an important pathway to drug discovery.

Complexes of Metals

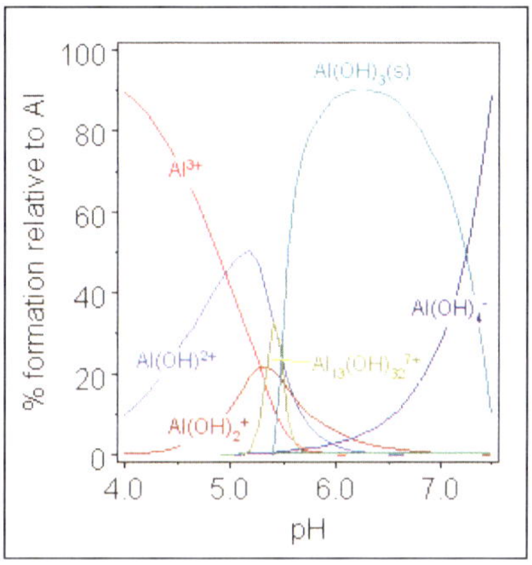

Speciation diagram for aluminium in aqueous solution as a function of pH. A variety of hydroxo complexes are formed, including aluminium hydroxide, ($Al(OH)_3$(s), which is insoluble at pH ~6.5.

The formation of a complex between a metal ion, M, and a ligand, L, is in fact usually a substitution reaction. For example, In aqueous solutions, metal ions will be present as aquo ions, so the reaction for the formation of the first complex could be written as:

$$[M(H_2O)_n] + L \rightleftharpoons [M(H_2O)_{n-1}L] + H_2O$$

However, since water is in vast excess, the concentration of water is usually assumed to be constant and is omitted from equilibrium constant expressions. Often, the metal and the ligand are in competition for protons. For the equilibrium:

$$p\,M + q\,L + r\,H \rightleftharpoons M_pL_qH_r$$

a stability constant can be defined as follows:

$$\beta_{pqr} = \frac{[M_p L_q H_r]}{[M]^p[L]^q[H]^r}$$

The definition can easily be extended to include any number of reagents. It includes hydroxide complexes because the concentration of the hydroxide ions is related to the concentration of hydrogen ions by the self-ionization of water:

$$[OH^-] = \frac{K_w}{[H^+]}$$

Stability constants defined in this way, are *association* constants. This can lead to some confusion as pK_a values are *dissociation* constants. In general purpose computer programs it is customary to define all constants as association constants. The relationship between the two types of constant is given in association and dissociation constants.

In biochemistry, an oxygen molecule can bind to an iron(II) atom in a heme prosthetic group in hemoglobin. The equilibrium is usually written, denoting hemoglobin by Hb, as:

$$Hb + O_2 \rightleftharpoons HbO_2$$

but this representation is incomplete as the Bohr effect shows that the equilibrium concentrations are pH-dependent. A better representation would be:

$$[HbH]^+ + O_2 \rightleftharpoons HbO_2 + H^+$$

as this shows that when hydrogen ion concentration increases the equilibrium is shifted to the left in accordance with Le Châtelier's principle. Hydrogen ion concentration can be increased by the presence of carbon dioxide, which behaves as a weak acid.

$$H_2O + CO_2 \rightleftharpoons HCO^-_3 + H^+$$

The iron atom can also bind to other molecules such as carbon monoxide. Cigarette smoke contains some carbon monoxide so the equilibrium:

$$HbO_2 + CO \rightleftharpoons Hb(CO) + O_2$$

is established in the blood of cigarette smokers.

Chelation therapy is based on the principle of using chelating ligands with a high binding selectivity for a particular metal to remove that metal from the human body.

Complexes with polyamino carboxylic acids find a wide range of applications. EDTA in particular is used extensively.

Redox Equilibria

A reduction–oxidation (redox) equilibrium can be handled in exactly the same way as any other chemical equilibrium. For example,

$$Fe^{2+} + Ce^{4+} \rightleftharpoons Fe^{3+} + Ce^{3+};\ K = \frac{[Fe^{3+}][Ce^{3+}]}{[Fe^{2+}][Ce^{4+}]}$$

However, in the case of redox reactions it is convenient to split the overall reaction into two half-reactions. In this example:

$$Fe^{3+} + e^- \rightleftharpoons Fe^{2+}$$

$$Ce^{4+} + e^- \rightleftharpoons Ce^{3+}$$

The standard free energy change, which is related to the equilibrium constant by:

$$\Delta G^{\ominus} = -RT \ln K$$

can be split into two components,

$$\Delta G^{\ominus} = \Delta G^{\ominus}_{\text{Fe}} + \Delta G^{\ominus}_{\text{Ce}}$$

The concentration of free electrons is effectively zero as the electrons are transferred directly from the reductant to the oxidant. The standard electrode potential, E^{o} for the each half-reaction is related to the standard free energy change by:

$$\Delta G^{\ominus}_{\text{Fe}} = -nFE^{0}_{\text{Fe}}; \Delta G^{\ominus}_{\text{Ce}} = -nFE^{0}_{\text{Ce}}$$

where n is the number of electrons transferred and F is the Faraday constant. Now, the free energy for an actual reaction is given by:

$$\Delta G = \Delta G^{\ominus} + RT \ln Q$$

where R is the gas constant and Q a reaction quotient. Strictly speaking Q is a quotient of activities, but it is common practice to use concentrations instead of activities. Therefore:

$$E_{\text{Fe}} = E^{0}_{\text{Fe}} + \frac{RT}{nF} \ln \frac{[\text{Fe}^{3+}]}{[\text{Fe}^{2+}]}$$

For any half-reaction, the redox potential of an actual mixture is given by the generalized expression:

$$E = E^{0} + \frac{RT}{nF} \ln \frac{[\text{oxidized species}]}{[\text{reduced species}]}$$

This is an example of the Nernst equation. The potential is known as a reduction potential. Standard electrode potentials are available in a table of values. Using these values, the actual electrode potential for a redox couple can be calculated as a function of the ratio of concentrations.

The equilibrium potential for a general redox half-reaction:

$$\alpha\,\text{A} + \beta\,\text{B}... + n\,\text{e}^{-} \rightleftharpoons \sigma\,\text{S} + \tau\,\text{T}...$$

is given by:

$$E = E^{\ominus} + \frac{RT}{nF} \ln \frac{\{\text{S}\}^{\sigma}\{\text{T}\}^{\tau}...}{\{\text{A}\}^{\alpha}\{\text{B}\}^{\beta}...}$$

Use of this expression allows the effect of a species not involved in the redox reaction, such as the hydrogen ion in a half-reaction such as:

$$MnO_4^- + 8\,H^+ + 5\,e^- \rightleftharpoons Mn^{2+} + 4\,H_2O$$

to be taken into account.

The equilibrium constant for a full redox reaction can be obtained from the standard redox potentials of the constituent half-reactions. At equilibrium the potential for the two half-reactions must be equal to each other and, of course, the number of electrons exchanged must be the same in the two half reactions.

Redox equilibria play an important role in the electron transport chain. The various cytochromes in the chain have different standard redox potentials, each one adapted for a specific redox reaction. This allows, for example, atmospheric oxygen to be reduced in photosynthesis. A distinct family of cytochromes, the cytochrome P450 oxidases, are involved in steroidogenesis and detoxification.

Solubility

When a solute forms a saturated solution in a solvent, the concentration of the solute, at a given temperature, is determined by the equilibrium constant at that temperature.

$$\ln K = -RT \ln\left(\frac{\sum_k a_k^{m_k}(\text{solution})}{a(\text{solid})}\right)$$

The activity of a pure substance in the solid state is one, by definition, so the expression simplifies to:

$$\ln K = -RT \ln\left(\sum_k a_k^{m_k}(\text{solution})\right)$$

If the solute does not dissociate the summation is replaced by a single term, but if dissociation occurs, as with ionic substances:

$$K_{SP} = \prod_k a_k^{m_k}$$

For example, with Na_2SO_4, $m_1 = 2$ and $m_2 = 1$ so the solubility product is written as:

$$K_{SP} = [\text{Na}^+]^2[\text{SO}_4^{2-}]$$

Concentrations, indicated by [...], are usually used in place of activities, but activity must be taken into account of the presence of another salt with no ions in common, the so-called salt effect. When another salt is present that has an ion in common, the common-ion effect comes into play, reducing the solubility of the primary solute.

Partition

When a solution of a substance in one solvent is brought into equilibrium with a second solvent that is immiscible with the first solvent, the dissolved substance may be partitioned between the two solvents. The ratio of concentrations in the two solvents is known as a partition coefficient or distribution coefficient. The partition coefficient is defined as the ratio of the analytical concentrations of the solute in the two phases. By convention the value is reported in logarithmic form.

$$\log p = \log \frac{[\text{solute}]_{\text{organic phase}}}{[\text{solute}]_{\text{aqueous phase}}}$$

The partition coefficient is defined at a specified temperature and, if applicable, pH of the aqueous phase. Partition coefficients are very important in pharmacology because they determine the extent to which a substance can pass from the blood (an aqueous solution) through a cell wall which is like an organic solvent. They are usually measured using water and octanol as the two solvents. Many pharmaceutical compounds are weak acids or weak bases. Such a compound may exist with a different extent of protonation depending on pH and the acid dissociation constant. Because the organic phase has a low dielectric constant the species with no electrical charge will be the most likely one to pass from the aqueous phase to the organic phase. Even at pH 7–7.2, the range of biological pH values, the aqueous phase may support an equilibrium between more than one protonated form. log *p* is determined from the analytical concentration of the substance in the aqueous phase, that is, the sum of the concentration of the different species in equilibrium.

An organic MTBE solution is extracted with aqueous sodium bicarbonate solution. This base removes benzoic acid as benzoate but leaves non-acidic benzil (yellow) behind in the upper organic phase.

Solvent extraction is used extensively in separation and purification processes. In its simplest form a reaction is performed in an organic solvent and unwanted by-products are removed by extraction into water at a particular pH.

A metal ion may be extracted from an aqueous phase into an organic phase in which the salt is not soluble, by adding a ligand. The ligand, L^{a-}, forms a complex with the metal ion, M^{b+}, $[ML_x]^{(b-ax)+}$ which has a strongly hydrophobic outer surface. If the complex has no electrical charge it will be extracted relatively easily into the organic phase. If the complex is charged, it is extracted as an ion pair. The additional ligand is not always required. For example, uranyl nitrate, $UO_2(NO_3)_2$, is soluble in diethyl ether because the solvent itself acts as a ligand. This property was used in the past for separating uranium from other metals whose salts are not soluble in ether. Currently extraction into kerosene is preferred, using a ligand such as tri-*n*-butyl phohsphate, TBP. In the PUREX process, which is commonly used in nuclear reprocessing, uranium(VI) is extracted from strong nitric acid as the electrically neutral complex $[UO_2(TBP)_2(NO_3)_2]$. The strong nitric acid provides a high concentration of nitrate ions which pushes the equilibrium in favour of the weak nitrato complex. Uranium is recovered by back-extraction (stripping) into weak nitric acid. Plutonium(IV) forms a similar complex, $[PuO_2(TBP)_2(NO_3)_2]$ and the plutonium in this complex can be reduced to separate it from uranium.

Another important application of solvent extraction is in the separation of the lanthanoids. This process also uses TBP and the complexes are extracted into kerosene. Separation is achieved because the stability constant for the formation of the TBP complex increases as the size of the lanthanoid ion decreases.

An instance of ion-pair extraction is in the use of a ligand to enable oxidation by potassium permanganate, $KMnO_4$, in an organic solvent. $KMnO_4$ is not soluble in organic solvents. When a ligand, such as a crown ether is added to an aqueous solution of $KMnO_4$, it forms a hydrophobic complex with the potassium cation which allows the uncharged ion pair $[KL]^+[MnO_4]^-$ to be extracted into the organic solvent. More complex partitioning problems (i.e. 3 or more phases present) can sometimes be handled with a fugacity capacity approach.

Chromatography

In chromatography substances are separated by partition between a stationary phase and a mobile phase. The analyte is dissolved in the mobile phase, and passes over the stationary phase. Separation occurs because of differing affinities of the analytes for the stationary phase. A distribution constant, K_d can be defined as:

$$K_d = \frac{a_s}{a_m}$$

where a_s and a_m are the equilibrium activities in the stationary and mobile phases respectively. It can be shown that the rate of migration, v, is related to the distribution constant by:

$$\bar{v} \propto \frac{1}{1 + fK_d}.$$

f is a factor which depends on the volumes of the two phases. Thus, the higher the affinity of the solute for the stationary phase, the slower the migration rate.

There is a wide variety of chromatographic techniques, depending on the nature of the stationary and mobile phases. When the stationary phase is solid, the analyte may form a complex with it. A water softener functions by selective complexation with a sulfonate ion exchange resin. Sodium ions form relatively weak complexes with the resin. When hard water is passed through the resin, the divalent ions of magnesium and calcium displace the sodium ions and are retained on the resin, *R*.

$$RNa + M^{2+} \rightleftharpoons RM^{+} + Na^{+}$$

The water coming out of the column is relatively rich in sodium ions and poor in calcium and magnesium which are retained on the column. The column is regenerated by passing a strong solution of sodium chloride through it, so that the resin–sodium complex is again formed on the column. Ion-exchange chromatography utilizes a resin such as chelex 100 in which iminodiacetate residues, attached to a polymer backbone, form chelate complexes of differing strengths with different metal ions, allowing the ions such as Cu^{2+} and Ni^{2+} to be separated chromatographically.

Another example of complex formation is in chiral chromatography in which is used to separate enantiomers from each other. The stationary phase is itself chiral and forms complexes selectively with the enantiomers. In other types of chromatography with a solid stationary phase, such as thin layer chromatography the analyte is selectively adsorbed onto the solid.

In gas–liquid chromatography (GLC) the stationary phase is a liquid such as polydimethylsiloxane, coated on a glass tube. Separation is achieved because the various components in the gas have different solubility in the stationary phase. GLC can be used to separate literally hundreds of components in a gas mixture such as cigarette smoke or essential oils, such as lavender oil.

Chemical Reaction

A chemical reaction is a process that leads to the chemical transformation of one set of chemical substances to another. Classically, chemical reactions encompass changes that only involve the positions of electrons in the forming and breaking of chemical bonds between atoms, with no change to the nuclei (no change to the elements present), and can often be described by a chemical equation. Nuclear chemistry is a sub-discipline of chemistry that involves the chemical reactions of unstable and radioactive elements where both electronic and nuclear changes can occur.

The substance (or substances) initially involved in a chemical reaction are called reactants or reagents. Chemical reactions are usually characterized by a chemical change, and they yield one or more products, which usually have properties different from the reactants. Reactions often consist of a sequence of individual sub-steps, the so-called elementary reactions, and the information on the precise course of action is part of the reaction mechanism. Chemical reactions are described with chemical equations, which symbolically present the starting materials, end products, and sometimes intermediate products and reaction conditions.

Chemical reactions happen at a characteristic reaction rate at a given temperature and chemical concentration. Typically, reaction rates increase with increasing temperature because there is more thermal energy available to reach the activation energy necessary for breaking bonds between atoms.

A thermite reaction using iron(III) oxide. The sparks flying outwards are globules of molten iron trailing smoke in their wake.

Reactions may proceed in the forward or reverse direction until they go to completion or reach equilibrium. Reactions that proceed in the forward direction to approach equilibrium are often described as spontaneous, requiring no input of free energy to go forward. Non-spontaneous reactions require input of free energy to go forward (examples include charging a battery by applying an external electrical power source, or photosynthesis driven by absorption of electromagnetic radiation in the form of sunlight).

Different chemical reactions are used in combinations during chemical synthesis in order to obtain a desired product. In biochemistry, a consecutive series of chemical reactions (where the product of one reaction is the reactant of the next reaction) form metabolic pathways. These reactions are often catalyzed by protein enzymes. Enzymes increase the rates of biochemical reactions, so that metabolic syntheses and decompositions impossible under ordinary conditions can occur at the temperatures and concentrations present within a cell.

The general concept of a chemical reaction has been extended to reactions between entities smaller than atoms, including nuclear reactions, radioactive decays, and reactions between elementary particles, as described by quantum field theory.

Chemical equations are used to graphically illustrate chemical reactions. They consist of chemical or structural formulas of the reactants on the left and those of the products on the right. They are separated by an arrow (→) which indicates the direction and type of the reaction; the arrow is read as the word "yields". The tip of the arrow points in the direction in which the reaction proceeds. A double arrow (⇌) pointing in opposite directions is used for equilibrium reactions. Equations should be balanced according to the stoichiometry, the number of atoms of each species should be the same on both sides of the equation. This is achieved by scaling the number of involved molecules ((A,B,C and D in a schematic example below) by the appropriate integers *a, b, c* and *d*.

$$aA + bB \rightarrow cC + dD$$

More elaborate reactions are represented by reaction schemes, which in addition to starting materials and products show important intermediates or transition states. Also, some relatively minor additions to the reaction can be indicated above the reaction arrow; examples of such additions are water, heat, illumination, a catalyst, etc. Similarly, some minor products can be placed below the arrow, often with a minus sign.

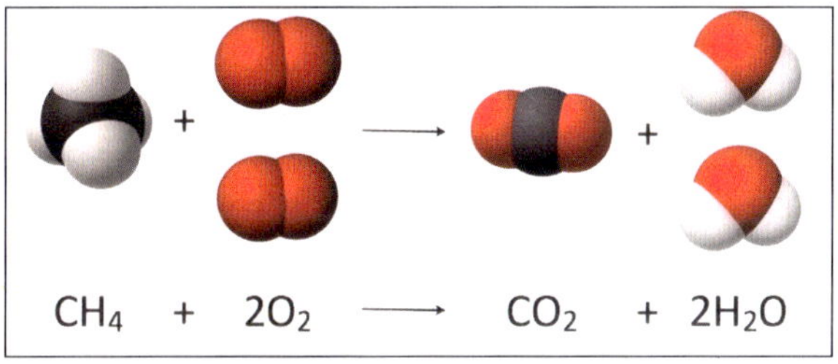

As seen from the equation $CH_4 + 2\,O_2 \rightarrow CO_2 + 2\,H_2O$, a coefficient of 2 must be placed before the oxygen gas on the reactants side and before the water on the products side in order for, as per the law of conservation of mass, the quantity of each element does not change during the reaction.

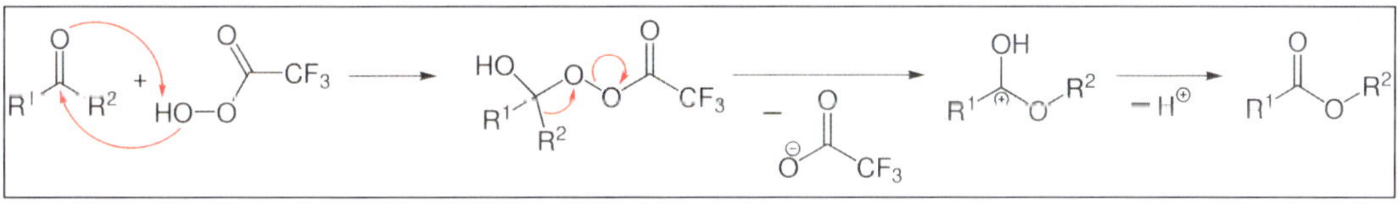

An example of organic reaction: oxidation of ketones to esters with a peroxycarboxylic acid.

Retrosynthetic analysis can be applied to design a complex synthesis reaction. Here the analysis starts from the products, for example by splitting selected chemical bonds, to arrive at plausible initial reagents. A special arrow (⇒) is used in retro reactions.

Elementary Reactions

The elementary reaction is the smallest division into which a chemical reaction can be decomposed, it has no intermediate products. Most experimentally observed reactions are built up from many elementary reactions that occur in parallel or sequentially. The actual sequence of the individual elementary reactions is known as reaction mechanism. An elementary reaction involves a few molecules, usually one or two, because of the low probability for several molecules to meet at a certain time.

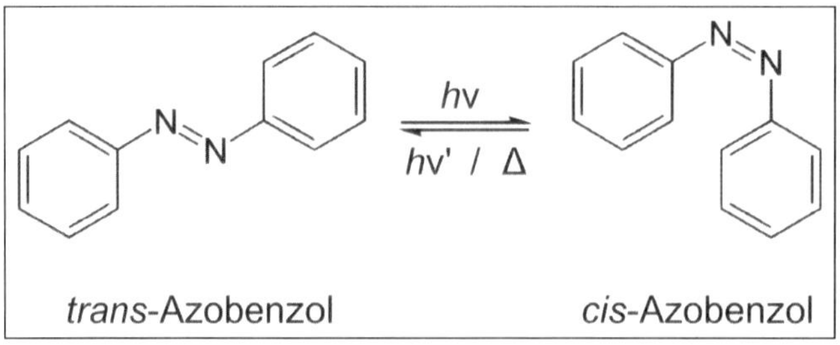

Isomerization of azobenzene, induced by light (hv) or heat (Δ).

The most important elementary reactions are unimolecular and bimolecular reactions. Only one molecule is involved in a unimolecular reaction; it is transformed by an isomerization or a dissociation into one or more other molecules. Such reactions require the addition of energy in the form of heat or light. A typical example of a unimolecular reaction is the cis–trans isomerization, in which the cis-form of a compound converts to the trans-form or vice versa.

In a typical dissociation reaction, a bond in a molecule splits (ruptures) resulting in two molecular fragments. The splitting can be homolytic or heterolytic. In the first case, the bond is divided so that each product retains an electron and becomes a neutral radical. In the second case, both electrons of the chemical bond remain with one of the products, resulting in charged ions. Dissociation plays an important role in triggering chain reactions, such as hydrogen–oxygen or polymerization reactions.

$$AB \rightarrow A + B$$

Dissociation of a molecule AB into fragments A and B.

For bimolecular reactions, two molecules collide and react with each other. Their merger is called chemical synthesis or an addition reaction.

$$A + B \rightarrow AB$$

Another possibility is that only a portion of one molecule is transferred to the other molecule. This type of reaction occurs, for example, in redox and acid-base reactions. In redox reactions, the transferred particle is an electron, whereas in acid-base reactions it is a proton. This type of reaction is also called metathesis.

$$HA + B \rightarrow A + HB$$

for example:

$$NaCl + AgNO_3 \rightarrow NaNO_3 + AgCl \downarrow$$

Chemical Equilibrium

Most chemical reactions are reversible, that is they can and do run in both directions. The forward and reverse reactions are competing with each other and differ in reaction rates. These rates depend on the concentration and therefore change with time of the reaction: the reverse rate gradually increases and becomes equal to the rate of the forward reaction, establishing the so-called chemical equilibrium. The time to reach equilibrium depends on such parameters as temperature, pressure and the materials involved, and is determined by the minimum free energy. In equilibrium, the Gibbs free energy must be zero. The pressure dependence can be explained with the Le Chatelier's principle. For example, an increase in pressure due to decreasing volume causes the reaction to shift to the side with the fewer moles of gas.

The reaction yield stabilizes at equilibrium, but can be increased by removing the product from the reaction mixture or changed by increasing the temperature or pressure. A change in the concentrations of the reactants does not affect the equilibrium constant, but does affect the equilibrium position.

Thermodynamics

Chemical reactions are determined by the laws of thermodynamics. Reactions can proceed by themselves if they are exergonic, that is if they release energy. The associated free energy of the reaction is composed of two different thermodynamic quantities, enthalpy and entropy:

$$\Delta G = \Delta H - T \cdot \Delta S.$$

G: free energy, *H*: enthalpy, *T*: temperature, *S*: entropy, Δ: difference(change between original and product).

Reactions can be exothermic, where ΔH is negative and energy is released. Typical examples of exothermic reactions are precipitation and crystallization, in which ordered solids are formed from disordered gaseous or liquid phases. In contrast, in endothermic reactions, heat is consumed from the environment. This can occur by increasing the entropy of the system, often through the formation of gaseous reaction products, which have high entropy. Since the entropy increases with temperature, many endothermic reactions preferably take place at high temperatures. On the contrary, many exothermic reactions such as crystallization occur at low temperatures. Changes in temperature can sometimes reverse the sign of the enthalpy of a reaction, as for the carbon monoxide reduction of molybdenum dioxide:

$$2\mathrm{CO(g)} + \mathrm{MoO_2(s)} \rightarrow 2\mathrm{CO_2(g)} + \mathrm{Mo(s)};$$
$$\Delta H^o = +21.86\ \mathrm{kJ\ \ at\ \ 298\ K}$$

This reaction to form carbon dioxide and molybdenum is endothermic at low temperatures, becoming less so with increasing temperature. ΔH° is zero at 1855 K, and the reaction becomes exothermic above that temperature.

Changes in temperature can also reverse the direction tendency of a reaction. For example, the water gas shift reaction:

$$CO(g) + H_2O(v) \rightleftharpoons CO_2(g) + H_2(g)$$

is favored by low temperatures, but its reverse is favored by high temperature. The shift in reaction direction tendency occurs at 1100 K.

Reactions can also be characterized by the internal energy which takes into account changes in the entropy, volume and chemical potential. The latter depends, among other things, on the activities of the involved substances.

$$dU = T \cdot dS - p \cdot dV + \mu \cdot dn$$

U: internal energy, S: entropy, p: pressure, μ: chemical potential, n: number of molecules, d: small change sign.

Kinetics

The speed at which reactions takes place is studied by reaction kinetics. The rate depends on various parameters, such as:

- Reactant concentrations, which usually make the reaction happen at a faster rate if raised through increased collisions per unit time. Some reactions, however, have rates that are *independent* of reactant concentrations. These are called zero order reactions.
- Surface area available for contact between the reactants, in particular solid ones in heterogeneous systems. Larger surface areas lead to higher reaction rates.
- Pressure – Increasing the pressure decreases the volume between molecules and therefore increases the frequency of collisions between the molecules.
- Activation energy, which is defined as the amount of energy required to make the reaction start and carry on spontaneously. Higher activation energy implies that the reactants need more energy to start than a reaction with a lower activation energy.
- Temperature, which hastens reactions if raised, since higher temperature increases the energy of the molecules, creating more collisions per unit time,
- The presence or absence of a catalyst. Catalysts are substances which change the pathway (mechanism) of a reaction which in turn increases the speed of a reaction by lowering the activation energy needed for the reaction to take place. A catalyst is not destroyed or changed during a reaction, so it can be used again.

- For some reactions, the presence of electromagnetic radiation, most notably ultraviolet light, is needed to promote the breaking of bonds to start the reaction. This is particularly true for reactions involving radicals.

Several theories allow calculating the reaction rates at the molecular level. This field is referred to as reaction dynamics. The rate υ of a first-order reaction, which could be disintegration of a substance A, is given by:

$$\upsilon = -\frac{d[\mathrm{A}]}{dt} = k \cdot [\mathrm{A}].$$

Its integration yields:

$$[\mathrm{A}](t) = [\mathrm{A}]_0 \cdot e^{-k \cdot t}.$$

Here k is first-order rate constant having dimension 1/time, [A](t) is concentration at a time *t* and $[A]_0$ is the initial concentration. The rate of a first-order reaction depends only on the concentration and the properties of the involved substance, and the reaction itself can be described with the characteristic half-life. More than one time constant is needed when describing reactions of higher order. The temperature dependence of the rate constant usually follows the Arrhenius equation:

$$k = k_0 \, e^{-E_a / k_B T}$$

where E_a is the activation energy and k_B is the Boltzmann constant. One of the simplest models of reaction rate is the collision theory. More realistic models are tailored to a specific problem and include the transition state theory, the calculation of the potential energy surface, the Marcus theory and the Rice–Ramsperger–Kassel–Marcus (RRKM) theory.

Reaction Types

Four Basic Types

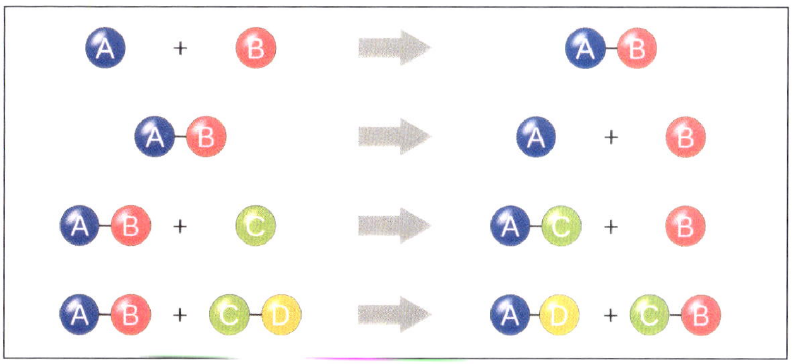

Representation of four basic chemical reactions types: Synthesis, decomposition, single replacement and double replacement.

Synthesis

In a synthesis reaction, two or more simple substances combine to form a more complex substance. These reactions are in the general form:

$$A + B \rightarrow AB$$

Two or more reactants yielding one product is another way to identify a synthesis reaction. One example of a synthesis reaction is the combination of iron and sulfur to form iron(II) sulfide:

$$8Fe + S_8 \rightarrow 8FeS$$

Another example is simple hydrogen gas combined with simple oxygen gas to produce a more complex substance, such as water.

Decomposition

A decomposition reaction is when a more complex substance breaks down into its more simple parts. It is thus the opposite of a synthesis reaction, and can be written as:

$$AB \rightarrow A + B$$

One example of a decomposition reaction is the electrolysis of water to make oxygen and hydrogen gas:

$$2H_2O \rightarrow 2H_2 + O_2$$

Single Replacement

In a single replacement reaction, a single uncombined element replaces another in a compound; in other words, one element trades places with another element in a compound These reactions come in the general form of:

$$A + BC \rightarrow AC + B$$

One example of a single displacement reaction is when magnesium replaces hydrogen in water to make magnesium hydroxide and hydrogen gas:

$$Mg + 2H_2O \rightarrow Mg(OH)_2 + H_2 \uparrow$$

Double Replacement

In a double replacement reaction, the anions and cations of two compounds switch places and form two entirely different compounds. These reactions are in the general form:

$$AB + CD \rightarrow AD + CB$$

For example, when barium chloride ($BaCl_2$) and magnesium sulfate ($MgSO_4$) react, the SO_4^{2-} anion switches places with the $2Cl^-$ anion, giving the compounds $BaSO_4$ and $MgCl_2$.

Another example of a double displacement reaction is the reaction of lead(II) nitrate with potassium iodide to form lead(II) iodide and potassium nitrate:

$$Pb(NO_3)_2 + 2KI \rightarrow PbI_2 \downarrow + 2KNO_3$$

Oxidation and Reduction

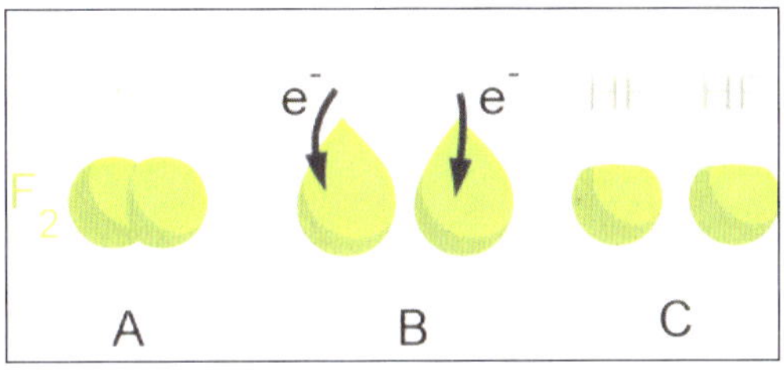

Illustration of a redox reaction.

Sodium chloride is formed through the redox reaction of sodium metal and chlorine gas.

Redox reactions can be understood in terms of transfer of electrons from one involved species (reducing agent) to another (oxidizing agent). In this process, the former species is *oxidized* and the latter is *reduced*. Though sufficient for many purposes, these descriptions are not precisely correct. Oxidation is better defined as an increase in oxidation state, and reduction as a decrease in oxidation state. In practice, the transfer of electrons will always change the oxidation state, but there are many reactions that are classed as "redox" even though no electron transfer occurs (such as those involving covalent bonds).

In the following redox reaction, hazardous sodium metal reacts with toxic chlorine gas to form the ionic compound sodium chloride, or common table salt:

$$2Na(s) + Cl_2(g) \rightarrow 2NaCl(s)$$

In the reaction, sodium metal goes from an oxidation state of 0 (as it is a pure element) to +1: in other words, the sodium lost one electron and is said to have been oxidized. On the other hand, the chlorine gas goes from an oxidation of 0 (it is also a pure element) to −1: the chlorine gains one electron and is said to have been reduced. Because the chlorine is the one reduced, it is considered the electron acceptor, or in other words, induces oxidation in the sodium – thus the chlorine gas is considered the oxidizing agent. Conversely, the sodium is oxidized or is the electron donor, and thus induces reduction in the other species and is considered the *reducing agent*.

Which of the involved reactants would be reducing or oxidizing agent can be predicted from the electronegativity of their elements. Elements with low electronegativity, such as most metals, easily donate electrons and oxidize – they are reducing agents. On the contrary, many ions with high oxidation numbers, such as H_2O_2, MnO_4^-, CrO_3, $Cr_2O_7^{2-}$, OsO_4 can gain one or two extra electrons and are strong oxidizing agents.

The number of electrons donated or accepted in a redox reaction can be predicted from the electron configuration of the reactant element. Elements try to reach the low-energy noble gas configuration, and therefore alkali metals and halogens will donate and accept one electron respectively. Noble gases themselves are chemically inactive.

An important class of redox reactions are the electrochemical reactions, where electrons from the power supply are used as the reducing agent. These reactions are particularly important for the production of chemical elements, such as chlorine or aluminium. The reverse process in which electrons are released in redox reactions and can be used as electrical energy is possible and used in batteries.

Complexation

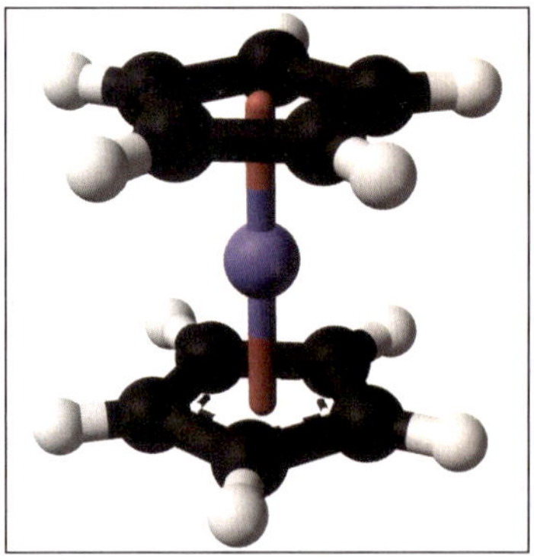

Ferrocene – an iron atom sandwiched between two C_5H_5 ligands.

In complexation reactions, several ligands react with a metal atom to form a coordination complex. This is achieved by providing lone pairs of the ligand into empty orbitals of the metal atom and forming dipolar bonds. The ligands are Lewis bases, they can be both ions and neutral molecules, such as carbon monoxide, ammonia or water. The number of ligands that react with a central metal atom can be found using the 18-electron rule, saying that the valence shells of a transition metal will collectively accommodate 18 electrons, whereas the symmetry of the resulting complex can be predicted with the crystal field theory and ligand field theory. Complexation reactions also include ligand exchange, in which one or more ligands are replaced by another, and redox processes which change the oxidation state of the central metal atom.

Acid-base Reactions

In the Brønsted–Lowry acid–base theory, an acid-base reaction involves a transfer of protons (H^+) from one species (the acid) to another (the base). When a proton is removed from an acid, the resulting species is termed that acid's conjugate base. When the proton is accepted by a base, the resulting species is termed that base's conjugate acid. In other words, acids act as proton donors and bases act as proton acceptors according to the following equation:

$$\underset{\text{acid}}{HA} + \underset{\text{base}}{B} \rightleftharpoons \underset{\text{conjugated base}}{A^-} + \underset{\text{conjugated acid}}{HB^+}$$

The reverse reaction is possible, and thus the acid/base and conjugated base/acid are always in equilibrium. The equilibrium is determined by the acid and base dissociation constants (K_a and K_b) of the involved substances. A special case of the acid-base reaction is the neutralization where an acid and a base, taken at exactly same amounts, form a neutral salt.

Acid-base reactions can have different definitions depending on the acid-base concept employed. Some of the most common are:

- Arrhenius definition: Acids dissociate in water releasing H_3O^+ ions; bases dissociate in water releasing OH^- ions.
- Brønsted-Lowry definition: Acids are proton (H^+) donors, bases are proton acceptors; this includes the Arrhenius definition.
- Lewis definition: Acids are electron-pair acceptors, bases are electron-pair donors; this includes the Brønsted-Lowry definition.

Precipitation

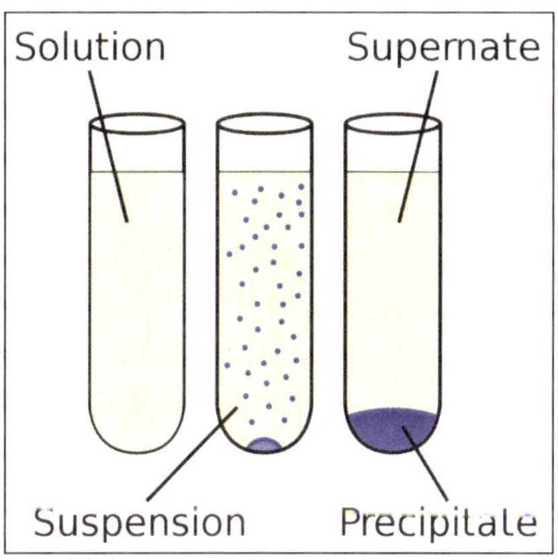

Precipitation

Precipitation is the formation of a solid in a solution or inside another solid during a chemical reaction. It usually takes place when the concentration of dissolved ions exceeds the solubility limit and forms an insoluble salt. This process can be assisted by adding a precipitating agent or by removal of the solvent. Rapid precipitation results in an amorphous or microcrystalline residue and slow process can yield single crystals. The latter can also be obtained by recrystallization from microcrystalline salts.

Solid-state Reactions

Reactions can take place between two solids. However, because of the relatively small diffusion rates in solids, the corresponding chemical reactions are very slow in comparison to liquid and gas phase reactions. They are accelerated by increasing the reaction temperature and finely dividing the reactant to increase the contacting surface area.

Reactions at the Solid-gas Interface

Reaction can take place at the solid-gas interface, surfaces at very low pressure such as ultra-high vacuum. Via scanning tunneling microscopy, it is possible to observe reactions at the solid|gas interface in real space, if the time scale of the reaction is in the correct range. Reactions at the solid|gas interface are in some cases related to catalysis.

Photochemical Reactions

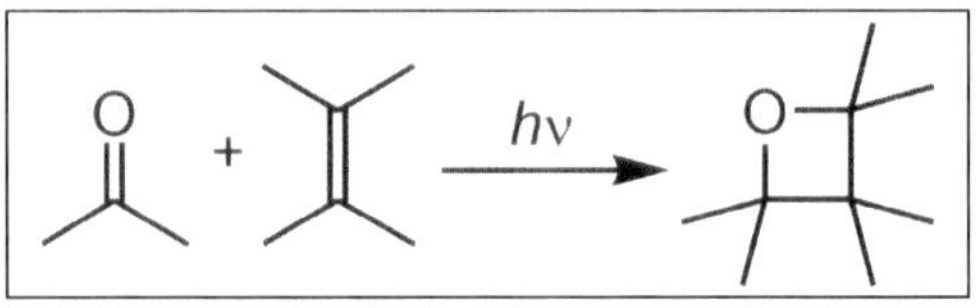

In this Paterno–Büchi reaction, a photoexcited carbonyl group is added to an unexcited olefin, yielding an oxetane.

In photochemical reactions, atoms and molecules absorb energy (photons) of the illumination light and convert into an excited state. They can then release this energy by breaking chemical bonds, thereby producing radicals. Photochemical reactions include hydrogen–oxygen reactions, radical polymerization, chain reactions and rearrangement reactions.

Many important processes involve photochemistry. The premier example is photosynthesis, in which most plants use solar energy to convert carbon dioxide and water into glucose, disposing of oxygen as a side-product. Humans rely on photochemistry for the formation of vitamin D, and vision is initiated by a photochemical reaction of rhodopsin. In fireflies, an enzyme in the abdomen catalyzes a reaction that results in bioluminescence. Many significant photochemical reactions, such as ozone formation, occur in the Earth atmosphere and constitute atmospheric chemistry.

Catalysis

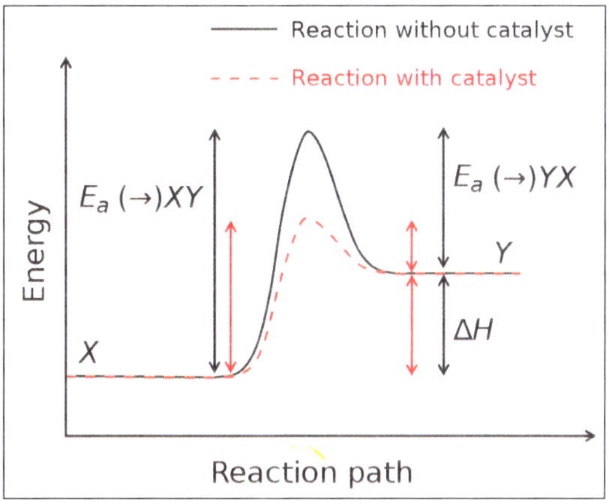

Schematic potential energy diagram showing the effect of a catalyst in an endothermic chemical reaction. The presence of a catalyst opens a different reaction pathway (in red) with a lower activation energy. The final result and the overall thermodynamics are the same.

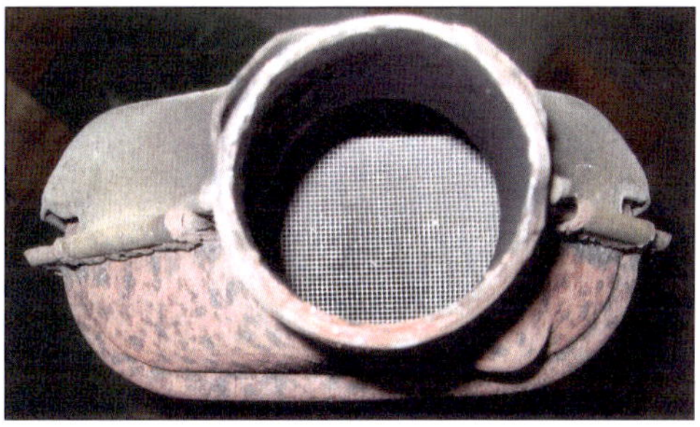

Solid heterogeneous catalysts are plated on meshes in ceramic catalytic converters in order to maximize their surface area. This exhaust converter is from a Peugeot 106 S2 1100.

In catalysis, the reaction does not proceed directly, but through reaction with a third substance known as catalyst. Although the catalyst takes part in the reaction, it is returned to its original state by the end of the reaction and so is not consumed. However, it can be inhibited, deactivated or destroyed by secondary processes. Catalysts can be used in a different phase (heterogeneous) or in the same phase (homogeneous) as the reactants. In heterogeneous catalysis, typical secondary processes include coking where the catalyst becomes covered by polymeric side products. Additionally, heterogeneous catalysts can dissolve into the solution in a solid–liquid system or evaporate in a solid–gas system. Catalysts can only speed up the reaction – chemicals that slow down the reaction are called inhibitors. Substances that increase the activity of catalysts are called promoters, and substances that deactivate catalysts are called catalytic poisons. With a catalyst, a reaction which is kinetically inhibited by a high activation energy can take place in circumvention of this activation energy.

Heterogeneous catalysts are usually solids, powdered in order to maximize their surface area. Of particular importance in heterogeneous catalysis are the platinum group metals and other transition metals, which are used in hydrogenations, catalytic reforming and in the synthesis of commodity chemicals such as nitric acid and ammonia. Acids are an example of a homogeneous catalyst, they increase the nucleophilicity of carbonyls, allowing a reaction that would not otherwise proceed with electrophiles. The advantage of homogeneous catalysts is the ease of mixing them with the reactants, but they may also be difficult to separate from the products. Therefore, heterogeneous catalysts are preferred in many industrial processes.

Reactions in Organic Chemistry

In organic chemistry, in addition to oxidation, reduction or acid-base reactions, a number of other reactions can take place which involve covalent bonds between carbon atoms or carbon and heteroatoms (such as oxygen, nitrogen, halogens, etc.). Many specific reactions in organic chemistry are name reactions designated after their discoverers.

Substitution

In a substitution reaction, a functional group in a particular chemical compound is replaced by another group. These reactions can be distinguished by the type of substituting species into a nucleophilic, electrophilic or radical substitution.

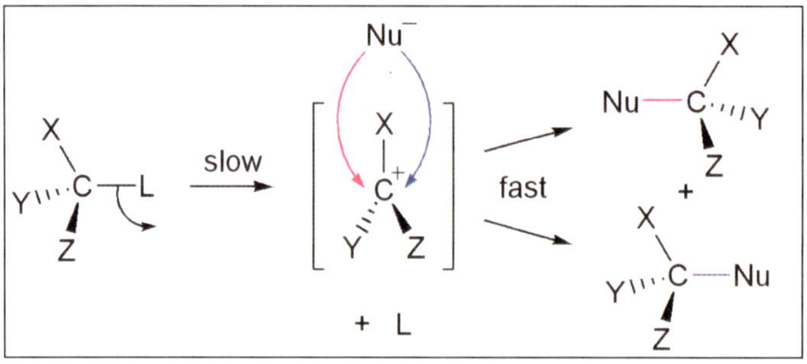

S_N1 mechanism

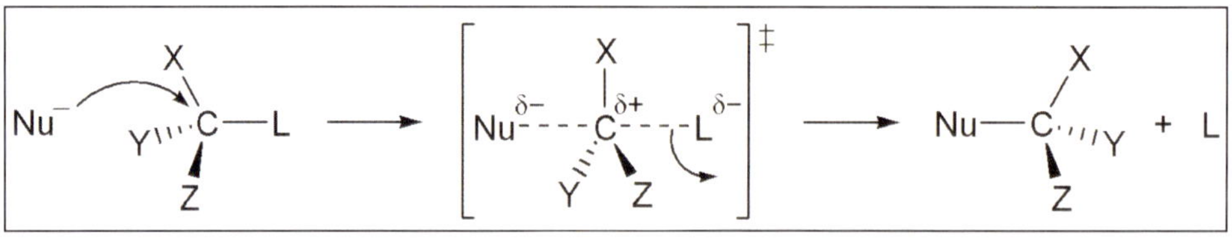

S_N2 mechanism

In the first type, a nucleophile, an atom or molecule with an excess of electrons and thus a negative charge or partial charge, replaces another atom or part of the "substrate" molecule. The electron pair from the nucleophile attacks the substrate forming a new bond, while the leaving group departs with an electron pair. The nucleophile may be electrically neutral or negatively charged, whereas the substrate is typically neutral or positively charged. Examples of nucleophiles are hydroxide ion, alkoxides, amines and halides. This type of reaction is found mainly in aliphatic hydrocarbons, and rarely in aromatic hydrocarbon. The latter have high electron density and enter nucleophilic aromatic substitution only with very strong electron withdrawing groups. Nucleophilic substitution can take place by two different mechanisms, S_N1 and S_N2. In their names, S stands for substitution, N for nucleophilic, and the number represents the kinetic order of the reaction, unimolecular or bimolecular.

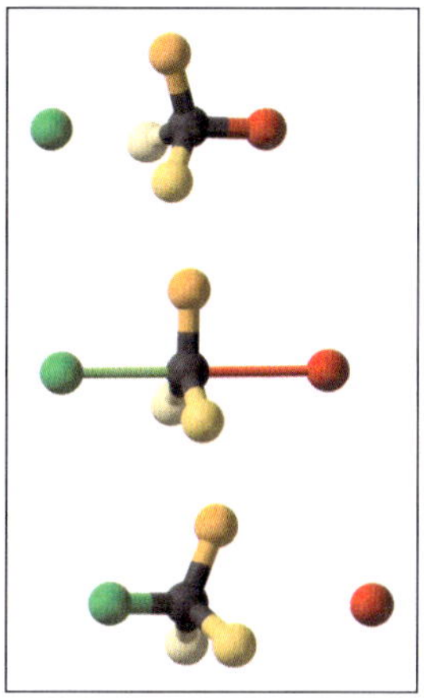

The three steps of an S_N2 reaction. The nucleophile is green and the leaving group is red.

S_N2 reaction causes stereo inversion (Walden inversion).

The S_N1 reaction proceeds in two steps. First, the leaving group is eliminated creating a carbocation. This is followed by a rapid reaction with the nucleophile.

In the S_N2 mechanism, the nucleophile forms a transition state with the attacked molecule, and only then the leaving group is cleaved. These two mechanisms differ in the stereochemistry of the products. S_N1 leads to the non-stereospecific addition and does not result in a chiral center, but rather in a set of geometric isomers (*cis/trans*). In contrast, a reversal (Walden inversion) of the previously existing stereochemistry is observed in the S_N2 mechanism.

Electrophilic substitution is the counterpart of the nucleophilic substitution in that the attacking atom or molecule, an electrophile, has low electron density and thus a positive charge. Typical electrophiles are the carbon atom of carbonyl groups, carbocations or sulfur or nitronium cations. This reaction takes place almost exclusively in aromatic hydrocarbons, where it is called electrophilic aromatic substitution. The electrophile attack results in the so-called σ-complex, a transition state in which the aromatic system is abolished. Then, the leaving group, usually a proton, is split off and the aromaticity is restored. An alternative to aromatic substitution is electrophilic aliphatic substitution. It is similar to the nucleophilic aliphatic substitution and also has two major types, S_E1 and S_E2.

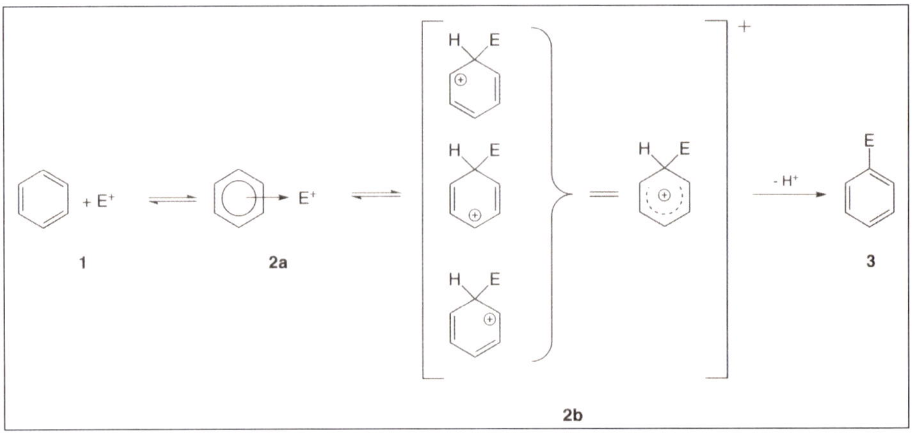

Mechanism of electrophilic aromatic substitution.

In the third type of substitution reaction, radical substitution, the attacking particle is a radical. This process usually takes the form of a chain reaction, for example in the reaction of alkanes with halogens. In the first step, light or heat disintegrates the halogen-containing molecules producing the radicals. Then the reaction proceeds as an avalanche until two radicals meet and recombine.

$$X\cdot + R-H \rightarrow X-H + R\cdot$$

$$R\cdot + X_2 \rightarrow R-X + X\cdot$$

Reactions during the chain reaction of radical substitution.

Addition and Elimination

The addition and its counterpart, the elimination, are reactions which change the number of substituents on the carbon atom, and form or cleave multiple bonds. Double and triple bonds can be produced by eliminating a suitable leaving group. Similar to the nucleophilic substitution, there are several possible reaction mechanisms which are named after the respective reaction order. In the E1 mechanism, the leaving group is ejected first, forming a carbocation. The next step, formation of the double bond, takes place with elimination of a proton (deprotonation). The leaving order is reversed in the E1cb mechanism, that is the proton is split off first. This mechanism requires participation of a base. Because of the similar conditions, both reactions in the E1 or E1cb elimination always compete with the S_N1 substitution.

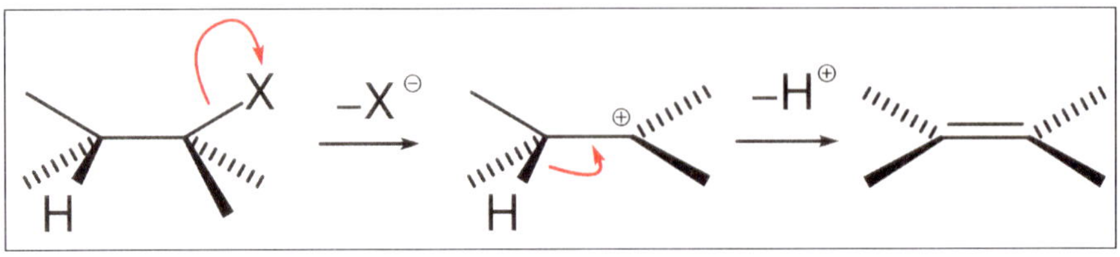

E1 elimination.

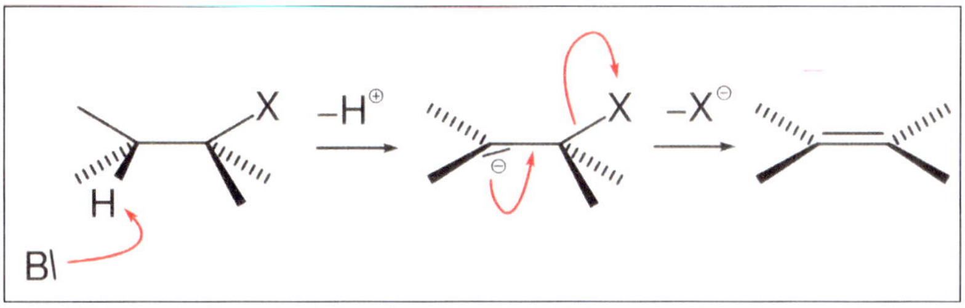

E1 cb elimination.

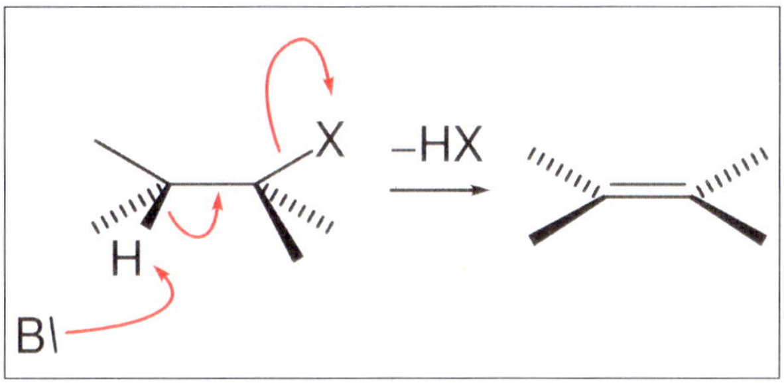

E2 elimination.

The E2 mechanism also requires a base, but there the attack of the base and the elimination of the leaving group proceed simultaneously and produce no ionic intermediate. In contrast to the E1 eliminations, different stereochemical configurations are possible for the reaction product in the E2 mechanism, because the attack of the base preferentially occurs in the anti-position with respect to the leaving group. Because of the similar conditions and reagents, the E2 elimination is always in competition with the S_N2-substitution.

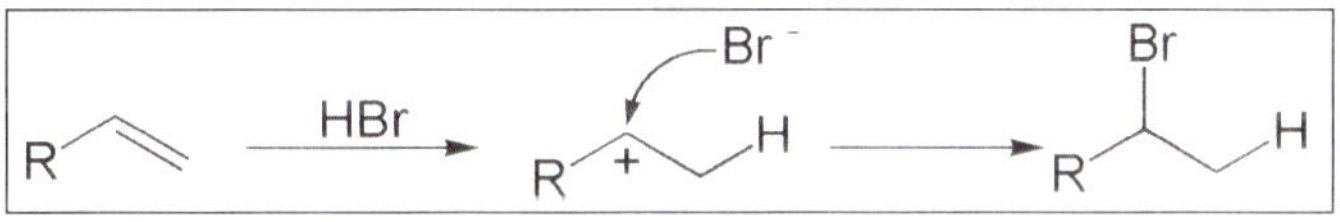

Electrophilic addition of hydrogen bromide.

The counterpart of elimination is the addition where double or triple bonds are converted into single bonds. Similar to the substitution reactions, there are several types of additions distinguished by the type of the attacking particle. For example, in the electrophilic addition of hydrogen bromide, an electrophile (proton) attacks the double bond forming a carbocation, which then reacts with the nucleophile (bromine). The carbocation can be formed on either side of the double bond depending on the groups attached to its ends, and the preferred configuration can be predicted with the Markovnikov's rule. This rule states that "In the heterolytic addition of a polar molecule to an alkene or alkyne, the more electronegative (nucleophilic) atom (or part) of the polar molecule becomes attached to the carbon atom bearing the smaller number of hydrogen atoms."

If the addition of a functional group takes place at the less substituted carbon atom of the double bond, then the electrophilic substitution with acids is not possible. In this case, one has to use the hydroboration–oxidation reaction, where in the first step, the boron atom acts as electrophile and adds to the less substituted carbon atom. At the second step, the nucleophilic hydroperoxide or halogen anion attacks the boron atom.

While the addition to the electron-rich alkenes and alkynes is mainly electrophilic, the nucleophilic addition plays an important role for the carbon-heteroatom multiple bonds, and especially its most important representative, the carbonyl group. This process is often associated with an elimination, so that after the reaction the carbonyl group is present again. It is therefore called addition-elimination reaction and may occur in carboxylic acid derivatives such as chlorides, esters or anhydrides. This reaction is often catalyzed by acids or bases, where the acids increase by the electrophilicity of the carbonyl group by binding to the oxygen atom, whereas the bases enhance the nucleophilicity of the attacking nucleophile.

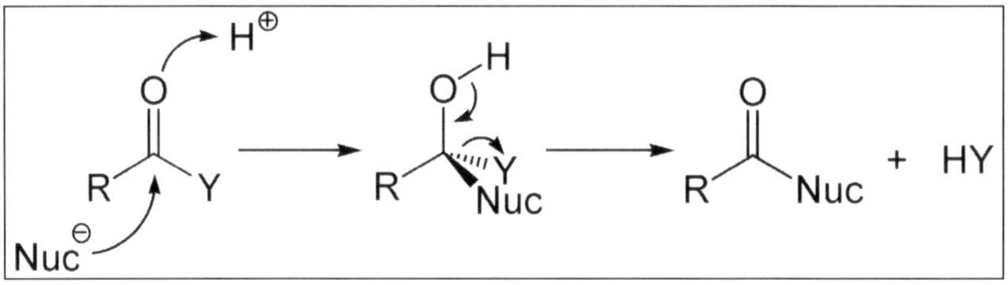

Acid-catalyzed addition-elimination mechanism.

Nucleophilic addition of a carbanion or another nucleophile to the double bond of an alpha, beta unsaturated carbonyl compound can proceed via the Michael reaction, which belongs to the larger class of conjugate additions. This is one of the most useful methods for the mild formation of C–C bonds.

Some additions which can not be executed with nucleophiles and electrophiles, can be succeeded with free radicals. As with the free-radical substitution, the radical addition proceeds as a chain reaction, and such reactions are the basis of the free-radical polymerization.

Other Organic Reaction Mechanisms

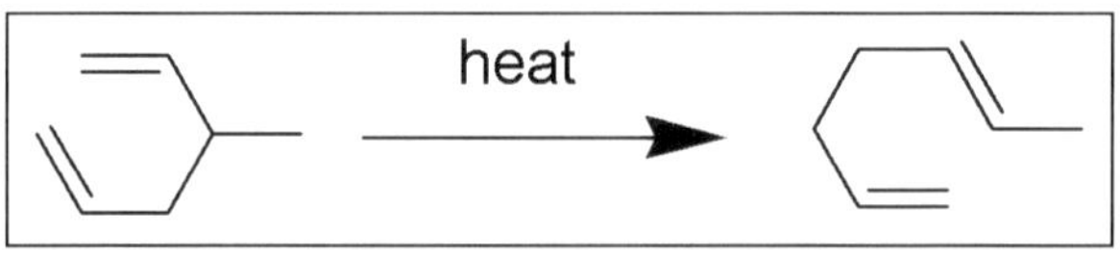

The Cope rearrangement of 3-methyl-1,5-hexadiene.

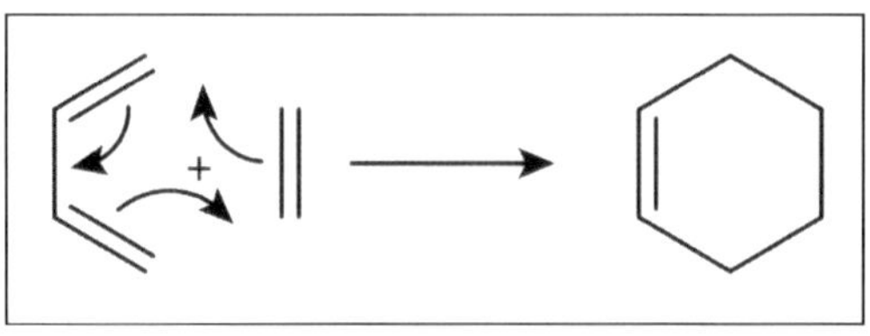

Mechanism of a Diels-Alder reaction.

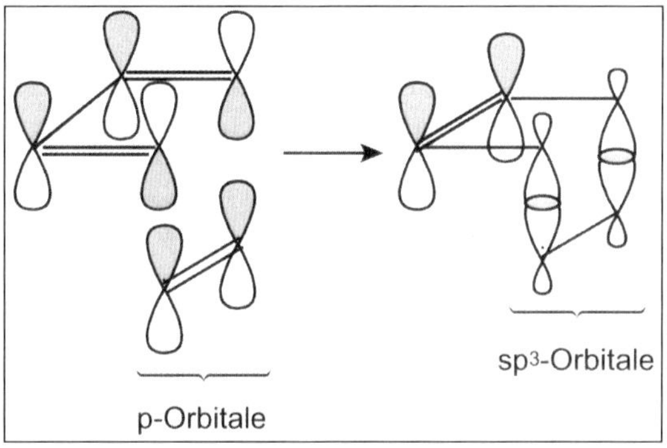

Orbital overlap in a Diels-Alder reaction.

In a rearrangement reaction, the carbon skeleton of a molecule is rearranged to give a structural isomer of the original molecule. These include hydride shift reactions such as the Wagner-Meerwein rearrangement, where a hydrogen, alkyl or aryl group migrates from one carbon to a neighboring carbon. Most rearrangements are associated with the breaking and formation of new carbon-carbon bonds. Other examples are sigmatropic reaction such as the Cope rearrangement.

Cyclic rearrangements include cycloadditions and, more generally, pericyclic reactions, wherein two or more double bond-containing molecules form a cyclic molecule. An important example of cycloaddition reaction is the Diels–Alder reaction (the so-called [4+2] cycloaddition) between a conjugated diene and a substituted alkene to form a substituted cyclohexene system.

Whether a certain cycloaddition would proceed depends on the electronic orbitals of the participating species, as only orbitals with the same sign of wave function will overlap and interact constructively to form new bonds. Cycloaddition is usually assisted by light or heat. These perturbations result in different arrangement of electrons in the excited state of the involved molecules and therefore in different effects. For example, the [4+2] Diels-Alder reactions can be assisted by heat whereas the [2+2] cycloaddition is selectively induced by light. Because of the orbital character, the potential for developing stereoisomeric products upon cycloaddition is limited, as described by the Woodward–Hoffmann rules.

Biochemical Reactions

Biochemical reactions are mainly controlled by enzymes. These proteins can specifically catalyze a single reaction, so that reactions can be controlled very precisely. The reaction takes place in the active site, a small part of the enzyme which is usually found in a cleft or pocket lined by amino acid residues, and the rest of the enzyme is used mainly for stabilization. The catalytic action of enzymes relies on several mechanisms including the molecular shape ("induced fit"), bond strain, proximity and orientation of molecules relative to the enzyme, proton donation or withdrawal (acid/base catalysis), electrostatic interactions and many others.

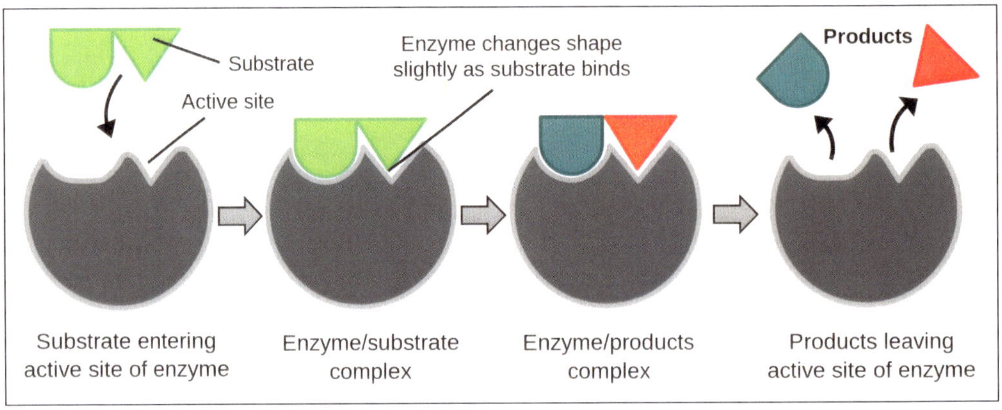

Illustration of the induced fit model of enzyme activity.

The biochemical reactions that occur in living organisms are collectively known as metabolism. Among the most important of its mechanisms is the anabolism, in which different DNA and enzyme-controlled processes result in the production of large molecules such as proteins and carbohydrates from smaller units. Bioenergetics studies the sources of energy for such reactions. An important energy source is glucose, which can be produced by plants via photosynthesis or assimilated from food. All organisms use this energy to produce adenosine triphosphate (ATP), which can then be used to energize other reactions.

Applications

Thermite reaction proceeding in railway welding. Shortly after this, the liquid iron flows into the mould around the rail gap.

Chemical reactions are central to chemical engineering where they are used for the synthesis of new compounds from natural raw materials such as petroleum and mineral ores. It is essential to make the reaction as efficient as possible, maximizing the yield and minimizing the amount of reagents, energy inputs and waste. Catalysts are especially helpful for reducing the energy required for the reaction and increasing its reaction rate.

Some specific reactions have their niche applications. For example, the thermite reaction is used to generate light and heat in pyrotechnics and welding. Although it is less controllable than the more conventional oxy-fuel welding, arc welding and flash welding, it requires much less equipment and is still used to mend rails, especially in remote areas.

Monitoring

Mechanisms of monitoring chemical reactions depend strongly on the reaction rate. Relatively slow processes can be analyzed in situ for the concentrations and identities of the individual ingredients. Important tools of real time analysis are the measurement of pH and analysis of optical absorption (color) and emission spectra. A less accessible but rather efficient method is introduction of a radioactive isotope into the reaction and monitoring how it changes over time and where it moves to; this method is often used to analyze redistribution of substances in the human body. Faster reactions are usually studied with ultrafast laser spectroscopy where utilization of femtosecond lasers allows short-lived transition states to be monitored at time scaled down to a few femtoseconds.

Reaction Rate

The reaction rate or rate of reaction is the speed at which reactants are converted into products. For example, the oxidative rusting of iron under Earth's atmosphere is a slow reaction that can take many years, but the combustion of cellulose in a fire is a reaction that takes place in fractions of a second. For most reactions, the rate decreases as the reaction proceeds.

Chemical kinetics is the part of physical chemistry that studies reaction rates. The concepts of chemical kinetics are applied in many disciplines, such as chemical engineering, enzymology and environmental engineering. Consider a typical chemical reaction:

$$a\mathrm{A} + b\mathrm{B} \rightarrow p\mathrm{P} + q\mathrm{Q}$$

The lowercase letters (a, b, p, and q) represent stoichiometric coefficients, while the capital letters represent the reactants (A and B) and the products (P and Q).

The reaction rate υ for a chemical reaction occurring in a closed system under isochoric conditions, without a build-up of reaction intermediates, is defined as:

$$\upsilon = -\frac{1}{a}\frac{d[\mathrm{A}]}{dt} = -\frac{1}{b}\frac{d[\mathrm{B}]}{dt} = \frac{1}{p}\frac{d[\mathrm{P}]}{dt} = \frac{1}{q}\frac{d[\mathrm{Q}]}{dt}$$

where [X] denotes the concentration of the substance X (= A, B, P or Q). Reaction rate thus defined has the units of mol/L/s.

The rate of a reaction is always positive. A negative sign is present to indicate that the reactant concentration is decreasing.) The IUPAC recommends that the unit of time should always be the second. The rate of reaction differs from the rate of increase of concentration of a product P by a constant factor (the reciprocal of its stoichiometric number) and for a reactant A by minus the reciprocal of the stoichiometric number. The stoichiometric numbers are included so that the defined rate is independent of which reactant or product species is chosen for measurement. For example, if $a = 1$ and $b = 3$ then B is consumed three times more rapidly than A, but $\upsilon = -d[A]/dt = -(1/3)d[B]/dt$ is uniquely defined. An additional advantage of this definition is that for an elementary and irreversible reaction, υ is equal to the product of probability of overcoming the transition state activation energy and the number of times per second the transition state is approached by reactant molecules. When so defined, for an elementary and irreversible reaction, υ is the rate of successful chemical reaction events leading to the product.

The above definition is only valid for a single reaction, in a closed system of constant volume, an assumption which should be stated explicitly in the definition. If water is added to a pot containing salty water, the concentration of salt decreases, although there is no chemical reaction.

For an open system, the full mass balance must be taken into account: in – out + generation – consumption = accumulation.

$$F_{A0} - F_A + \int_0^V \upsilon dV = \frac{dN_A}{dt},$$

where F_{Ao} is the inflow rate of A in molecules per second, F_A the outflow, and υ is the instantaneous reaction rate of A (in number concentration rather than molar) in a given differential volume, integrated over the entire system volume V at a given moment. When applied to the closed system at constant volume considered previously, this equation reduces to:

$$r = \frac{d[A]}{dt},$$

where the concentration [A] is related to the number of molecules N_A by $[A] = N_A/N_oV$. Here N_o is the Avogadro constant.

For a single reaction in a closed system of varying volume the so-called *rate of conversion* can be used, in order to avoid handling concentrations. It is defined as the derivative of the extent of reaction with respect to time.

$$\upsilon = \frac{d\xi}{dt} = \frac{1}{\nu_i}\frac{dn_i}{dt} = \frac{1}{\nu_i}\frac{d(C_iV)}{dt} = \frac{1}{\nu_i}\left(V\frac{dC_i}{dt} + C_i\frac{dV}{dt}\right)$$

Here ν_i is the stoichiometric coefficient for substance i, equal to a, b, p, and q in the typical reaction above. Also V is the volume of reaction and C_i is the concentration of substance i.

When side products or reaction intermediates are formed, the IUPAC recommends the use of the terms the rate of increase of concentration and rate of the decrease of concentration for products and reactants, properly.

Reaction rates may also be defined on a basis that is not the volume of the reactor. When a catalyst is used the reaction rate may be stated on a catalyst weight (mol g^{-1} s^{-1}) or surface area (mol m^{-2} s^{-1}) basis. If the basis is a specific catalyst site that may be rigorously counted by a specified method, the rate is given in units of s^{-1} and is called a turnover frequency.

Influencing Factors

Factors that influence the reaction rate are the nature of the reaction, concentration, pressure, reaction order, temperature, solvent, electromagnetic radiation, catalyst, isotopes, surface area, stirring, and diffusion limit. Some reactions are naturally faster than others. The number of reacting species, their physical state (the particles that form solids move much more slowly than those of gases or those in solution), the complexity of the reaction and other factors can greatly influence the rate of a reaction.

Reaction rate increases with concentration, as described by the rate law and explained by collision theory. As reactant concentration increases, the frequency of collision increases. The rate of gaseous reactions increases with pressure, which is, in fact, equivalent to an increase in concentration of the gas. The reaction rate increases in the direction where there are fewer moles of gas and decreases in the reverse direction. For condensed-phase reactions, the pressure dependence is weak.

The order of the reaction controls how the reactant concentration (or pressure) affects reaction rate.

Usually conducting a reaction at a higher temperature delivers more energy into the system and increases the reaction rate by causing more collisions between particles, as explained by collision theory. However, the main reason that temperature increases the rate of reaction is that more of the colliding particles will have the necessary activation energy resulting in more successful collisions (when bonds are formed between reactants). The influence of temperature is described by the Arrhenius equation. For example, coal burns in a fireplace in the presence of oxygen, but it does not when it is stored at room temperature. The reaction is spontaneous at low and high temperatures but at room temperature its rate is so slow that it is negligible. The increase in temperature, as created by a match, allows the reaction to start and then it heats itself, because it is exothermic. That is valid for many other fuels, such as methane, butane, and hydrogen.

Reaction rates can be independent of temperature (*non-Arrhenius*) or decrease with increasing temperature (*anti-Arrhenius*). Reactions without an activation barrier (e.g., some radical reactions), tend to have anti Arrhenius temperature dependence: the rate constant decreases with increasing temperature.

Many reactions take place in solution and the properties of the solvent affect the reaction rate. The ionic strength also has an effect on reaction rate.

Electromagnetic radiation is a form of energy. As such, it may speed up the rate or even make a reaction spontaneous as it provides the particles of the reactants with more energy. This energy is in one way or another stored in the reacting particles (it may break bonds, promote molecules to electronically or vibrationally excited states) creating intermediate species that react easily. As the intensity of light increases, the particles absorb more energy and hence the rate of reaction increases. For example, when methane reacts with chlorine in the dark, the reaction rate is slow. It can be sped up when the mixture is put under diffused light. In bright sunlight, the reaction is explosive.

The presence of a catalyst increases the reaction rate (in both the forward and reverse reactions) by providing an alternative pathway with a lower activation energy. For example, platinum catalyzes the combustion of hydrogen with oxygen at room temperature.

The kinetic isotope effect consists in a different reaction rate for the same molecule if it has different isotopes, usually hydrogen isotopes, because of the relative mass difference between hydrogen and deuterium. In reactions on surfaces, which take place for example during heterogeneous catalysis, the rate of reaction increases as the surface area does. That is because more particles of the solid are exposed and can be hit by reactant molecules. Stirring can have a strong effect on the rate of reaction for heterogeneous reactions.

Some reactions are limited by diffusion. All the factors that affect a reaction rate, except for concentration and reaction order, are taken into account in the reaction rate coefficient (the coefficient in the rate equation of the reaction).

Rate Equation

For a chemical reaction $a\,\mathrm{A} + b\,\mathrm{B} \rightarrow p\,\mathrm{P} + q\,\mathrm{Q}$, the rate equation or rate law is a mathematical expression used in chemical kinetics to link the rate of a reaction to the concentration of each reactant. It is often of the type:

$$v = k(T)[A]^n[B]^m$$

For gas phase reaction the rate is often alternatively expressed by partial pressures.

In these equations $k(T)$ is the *reaction rate coefficient* or *rate constant*, although it is not really a constant, because it includes all the parameters that affect reaction rate, except for concentration, which is explicitly taken into account. Of all the parameters influencing reaction rates, temperature is normally the most important one and is accounted for by the Arrhenius equation.

The exponents n and m are called reaction orders and depend on the reaction mechanism. For elementary (single-step) reactions the order with respect to each reactant is equal to its stoichiometric coefficient. For complex (multistep) reactions, however, this is often not true and the rate equation is determined by the detailed mechanism, as illustrated below for the reaction of H_2 and NO.

For elementary reactions or reaction steps, the order and stoichiometric coefficient are both equal to the molecularity or number of molecules participating. For a unimolecular reaction or step the rate is proportional to the concentration of molecules of reactant, so that the rate law is first order. For a bimolecular reaction or step, the number of collisions is proportional to the product of the two reactant concentrations, or second order. A termolecular step is predicted to be third order, but also very slow as simultaneous collisions of three molecules are rare.

By using the mass balance for the system in which the reaction occurs, an expression for the rate of change in concentration can be derived. For a closed system with constant volume, such an expression can look like:

$$\frac{d[P]}{dt} = k(T)[A]^n[B]^m$$

Example of a Complex Reaction: Hydrogen and Nitric Oxide

For the reaction $2H_2(g)+2NO(g) \rightarrow N_2(g)+2H_2O(g)$ the observed rate equation (or rate expression) is: $\upsilon = k[H_2][NO]^2$.

As for many reactions, the experimental rate equation does not simply reflect the stoichiometric coefficients in the overall reaction: It is third order overall: first order in H_2 and second order in NO, even though the stoichiometric coefficients of both reactants are equal to 2.

In chemical kinetics, the overall reaction rate is often explained using a mechanism consisting of a number of elementary steps. Not all of these steps affect the rate of reaction; normally the slowest elementary step controls the reaction rate. For this example, a possible mechanism is:

$$2NO(g) \rightleftharpoons N_2O_2(g) \quad \text{(fast equilibrium)}$$

$$N_2O_2 + H_2 \rightarrow N_2O + H_2O \quad \text{(slow)}$$

$$N_2O + H_2 \rightarrow N_2 + H_2O \quad \text{(fast)}$$

First and third reactions are very rapid compared to the second, so the slow reaction is the rate determining step. This is a bimolecular elementary reaction whose rate is given by the second order equation:

$$\upsilon = k_2[H_2][N_2O_2],$$

where k_2 is the rate constant for the second step.

However N_2O_2 is an unstable intermediate whose concentration is determined by the fact that the first step is in equilibrium, so that $[N_2O_2] = K_1[NO]^2$, where K_1 is the equilibrium constant of the first step. Substitution of this equation in the previous equation leads to a rate equation expressed in terms of the original reactants:

$$v = k_2 K_1 [H_2][NO]^2$$

This agrees with the form of the observed rate equation if it is assumed that $k = k_2K_1$. In practice the rate equation is used to suggest possible mechanisms which predict a rate equation in agreement with experiment.

The second molecule of H_2 does not appear in the rate equation because it reacts in the third step, which is a rapid step *after* the rate-determining step, so that it does not affect the overall reaction rate.

Temperature Dependence

Each reaction rate coefficient k has a temperature dependency, which is usually given by the Arrhenius equation:

$$k = Ae^{-\frac{E_a}{RT}}$$

E_a is the activation energy and R is the gas constant. Since at temperature T the molecules have energies given by a Boltzmann distribution, one can expect the number of collisions with energy greater than E_a to be proportional to $e^{-Ea/RT}$. A is the pre-exponential factor or frequency factor.

The values for A and E_a are dependent on the reaction. There are also more complex equations possible, which describe temperature dependence of other rate constants that do not follow this pattern.

A chemical reaction takes place only when the reacting particles collide. However, not all collisions are effective in causing the reaction. Products are formed only when the colliding particles possess a certain minimum energy called threshold energy. As a rule of thumb, reaction rates for many reactions double for every 10 degrees Celsius increase in temperature, For a given reaction, the ratio of its rate constant at a higher temperature to its rate constant at a lower temperature is known as its temperature coefficient (Q). Q_{10} is commonly used as the ratio of rate constants that are 10 °C apart.

Pressure Dependence

The pressure dependence of the rate constant for condensed-phase reactions (i.e., when reactants and products are solids or liquid) is usually sufficiently weak in the range of pressures normally encountered in industry that it is neglected in practice.

The pressure dependence of the rate constant is associated with the activation volume. For the reaction proceeding through an activation-state complex:

$$A + B \rightleftharpoons |A \cdots B|^{\ddagger} \rightarrow P$$

the activation volume, $\Delta V^{\ddagger}$, is:

$$\Delta V^{\ddagger} = \bar{V}_{\ddagger} - \bar{V}_A - \bar{V}_B$$

where $\bar{V}$ denotes the partial molar volume of a species and ‡ indicates the activation-state complex.

For the above reaction, one can expect the change of the reaction rate constant (based either on mole fraction or on molar concentration) with pressure at constant temperature to be:

$$\left(\frac{\partial \ln k_x}{\partial P}\right)_T = -\frac{\Delta V^{\ddagger}}{RT}$$

In practice, the matter can be complicated because the partial molar volumes and the activation volume can themselves be a function of pressure.

Reactions can increase or decrease their rates with pressure, depending on the value of $\Delta V^{\ddagger}$. As an example of the possible magnitude of the pressure effect, some organic reactions were shown to double the reaction rate when the pressure was increased from atmospheric (0.1 MPa) to 50 MPa (which gives $\Delta V^{\ddagger} = -0.025$ L/mol).

Rate Equation

The rate law or rate equation for a chemical reaction is an equation that links the reaction rate with the concentrations or pressures of the reactants and constant parameters (normally rate coefficients and partial reaction orders). For many reactions the rate is given by a power law such as:

$$v = k[A]^x[B]^y$$

where [A] and [B] express the concentration of the species A and B (usually in moles per liter (molarity, M)). The exponents x and y are the partial *orders of reaction* for A and B and the *overall* reaction order is the sum of the exponents. These are often positive integers, but they may also be zero, fractional, or negative. The constant k is the reaction rate constant or *rate coefficient* of the reaction. Its value may depend on conditions such as temperature, ionic strength, surface area of an adsorbent, or light irradiation.

Elementary (single-step) reactions and reaction steps have reaction orders equal to the stoichiometric coefficients for each reactant. The overall reaction order, i.e. the sum of stoichiometric coefficients of reactants, is always equal to the molecularity of the elementary reaction. However complex (multi-step) reactions may or may not have reaction orders equal to their stoichiometric coefficients. This implies that the order and the rate equation of a given reaction cannot be reliably deduced from the stoichiometry and must be determined experimentally, since an unknown reaction mechanism could be either elementary or complex. When the experimental rate equation has been determined, it is often of use for deduction of the reaction mechanism.

The rate equation of a reaction with an assumed multi-step mechanism can often be derived theoretically using quasi-steady state assumptions from the underlying elementary reactions, and compared with the experimental rate equation as a test of the assumed mechanism. The equation may involve a fractional order, and may depend on the concentration of an intermediate species.

A reaction can also have an *undefined* reaction order with respect to a reactant if the rate is not simply proportional to some power of the concentration of that reactant; for example, one cannot talk about reaction order in the rate equation for a bimolecular reaction between adsorbed molecules:

$$\upsilon = k\frac{K_1 K_2 C_A C_B}{(1+K_1 C_A + K_2 C_B)^2}.$$

Consider a typical chemical reaction in which two reactants A and B combine to form a product C:

$$\mathrm{A} + 2\mathrm{B} \rightarrow 3\mathrm{C}.$$

This can also be written:

$$0 = -\mathrm{A} - 2\mathrm{B} + 3\mathrm{C}.$$

The prefactors -1, -2 and 3 (with negative signs for reactants because they are consumed) are known as stoichiometric coefficients. One molecule of A combines with two of B to form 3 of C, so if we use the symbol [X] for the number of moles of chemical X,

$$-\frac{d[\mathrm{A}]}{dt} = -\frac{1}{2}\frac{d[\mathrm{B}]}{dt} = \frac{1}{3}\frac{d[\mathrm{C}]}{dt}.$$

If the reaction takes place in a closed system at constant temperature and volume, without a build-up of reaction intermediates, the *reaction rate r* is defined as:

$$\upsilon = \frac{1}{\nu_i}\frac{d[\mathrm{X}_i]}{dt},$$

where ν_i is the stoichiometric coefficient for chemical X_i.

The reaction rate has some functional dependence on the concentrations of the reactants,

$$v = f([\mathrm{A}],[\mathrm{B}],\ldots),$$

and this dependence is known as the *rate equation* or *rate law*. This law generally cannot be deduced from the chemical equation and must be determined by experiment.

Power Laws

A common form for the rate equation is a power law:

$$v = k[\mathrm{A}]^x[\mathrm{B}]^y\ldots$$

The constant k is called the *rate constant*. The exponents, which can be fractional, are called *partial orders of reaction* and their sum is the overall order of reaction.

In a dilute solution, an elementary reaction (one having a single step with a single transition state) is empirically found to obey the law of mass action. This predicts that the rate depends only on the concentrations of the reactants, raised to the powers of their stoichiometric coefficients.

Determination of Reaction Order

Method of Initial Rates

The natural logarithm of the power-law rate equation is:

$$\ln v = \ln k + x\ln[A] + y\ln[B] + \ldots$$

This can be used to estimate the order of reaction of each reactant. For example, the initial rate can be measured in a series of experiments at different initial concentrations of reactant A with all other concentrations [B], [C], ... kept constant, so that:

$$\ln v = x\ln[A] + \text{constant}.$$

The slope of a graph of $\ln v$ as a function of $\ln[A]$ then corresponds to the order x with respect to reactant A.

However, this method is not always reliable because:

1. Measurement of the initial rate requires accurate determination of small changes in concentration in short times (compared to the reaction half-life) and is sensitive to errors.

2. The rate equation will not be completely determined if the rate also depends on substances not present at the beginning of the reaction, such as intermediates or products.

Integral Method

The tentative rate equation determined by the method of initial rates is therefore normally verified by comparing the concentrations measured over a longer time (several half-lives) with the integrated form of the rate equation.

For example, the integrated rate law for a first-order reaction is:

$$\ln[A] = -kt + \ln[A]_0,$$

where [A] is the concentration at time t and $[A]_0$ is the initial concentration at zero time. The first-order rate law is confirmed if $\ln[A]$ is in fact a linear function of time. In this case the rate constant k is equal to the slope with sign reversed.

Method of Flooding

The partial order with respect to a given reactant can be evaluated by the method of flooding (or of isolation) of Ostwald. In this method, the concentration of one reactant is measured with all other reactants in large excess so that their concentration remains essentially constant. For a reaction a·A + b·B → c·C with rate law: $\upsilon = k \cdot [\mathrm{A}]^x \cdot [\mathrm{B}]^y$, the partial order x with respect to A is determined using a large excess of B. In this case:

$$\upsilon = k' \cdot [\mathrm{A}]^x \text{ with } k' = k \cdot [\mathrm{B}]^y k' = k \cdot [\mathrm{B}]^y,$$

and x may be determined by the integral method. The order y with respect to B under the same conditions (with B in excess) is determined by a series of similar experiments with a range of initial concentration $[\mathrm{B}]_0$ so that the variation of k' can be measured.

Zero Order

For zero-order reactions, the reaction rate is independent of the concentration of a reactant, so that changing its concentration has no effect on the speed of the reaction. Thus, the concentration changes linearly with time. This may occur when there is a bottleneck which limits the number of reactant molecules that can react at the same time, for example if the reaction requires contact with an enzyme or a catalytic surface.

Many enzyme-catalyzed reactions are zero order, provided that the reactant concentration is much greater than the enzyme concentration which controls the rate, so that the enzyme is *saturated*. For example, the biological oxidation of ethanol to acetaldehyde by the enzyme liver alcohol dehydrogenase (LADH) is zero order in ethanol.

Similarly reactions with heterogeneous catalysis can be zero order if the catalytic surface is saturated. For example, the decomposition of phosphine (PH_3) on a hot tungsten surface at high pressure is zero order in phosphine which decomposes at a constant rate.

In homogeneous catalysis zero order behavior can come about from reversible inhibition. For example, ring-opening metathesis polymerization using third-generation Grubbs catalyst exhibits zero order behavior in catalyst due to the reversible inhibition that is occur between the pyridine and the ruthenium center.

First Order

A *first order reaction* depends on the concentration of only one reactant (a *unimolecular reaction*). Other reactants can be present, but each will be zero order. The rate law for such a reaction is:

$$-\frac{d[\mathrm{A}]}{dt} = k[\mathrm{A}],$$

The half-life is independent of the starting concentration and is given by $t_{1/2} = \frac{\ln(2)}{k}$.

Examples of such reactions are:

$$\mathrm{H_2O_2(l) \rightarrow H_2O(l) + \frac{1}{2}O_2(g)}$$

$$\mathrm{SO_2Cl_2(l) \rightarrow SO_2(g) + Cl_2(g)}$$

$$\mathrm{2N_2O_5(g) \rightarrow 4NO_2(g) + O_2(g)}$$

In organic chemistry, the class of S_N1 (nucleophilic substitution unimolecular) reactions consists of first-order reactions. For example, in the reaction of aryldiazonium ions with nucleophiles in aqueous solution $ArN_2^+ + X^- \rightarrow ArX + N_2$, the rate equation is $v = k[ArN_2^+]$, where Ar indicates an aryl group.

Second Order

A reaction is said to be second order when the overall order is two. The rate of a second-order reaction may be proportional to one concentration squared $v = k[A]^2$, or (more commonly) to the product of two concentrations $v = k[A][B]$. As an example of the first type, the reaction $NO_2 + CO \rightarrow NO + CO_2$ is second-order in the reactant NO_2 and zero order in the reactant CO. The observed rate is given by $v = k[NO_2]^2$, and is independent of the concentration of CO.

For the rate proportional to a single concentration squared, the time dependence of the concentration is given by:

$$\frac{1}{[\mathrm{A}]} = \frac{1}{[\mathrm{A}]_0} + kt$$

The time dependence for a rate proportional to two unequal concentrations is:

$$\frac{[A]}{[B]}=\frac{[A]_0}{[B]_0}e^{([A]_0-[B]_0)kt};$$

if the concentrations are equal, they satisfy the previous equation.

The second type includes nucleophillic addition-elimination reactions, such as the alkaline hydrolysis of ethyl acetate:

$$CH_3COOC_2H_5 + OH^- \rightarrow CH_3COO^- + C_2H_5OH$$

This reaction is first-order in each reactant and second-order overall:

$$r = k[CH_3COOC_2H_5][OH^-]$$

If the same hydrolysis reaction is catalyzed by imidazole, the rate equation becomes $\upsilon = k[\text{imidazole}][CH_3COOC_2H_5]$. The rate is first-order in one reactant (ethyl acetate), and also first-order in imidazole which as a catalyst does not appear in the overall chemical equation.

Another well-known class of second-order reactions are the S_N2 (bimolecular nucleophilic substitution) reactions, such as the reaction of n-butyl bromide with sodium iodide in acetone:

$$CH_3CH_2CH_2CH_2Br + NaI \rightarrow CH_3CH_2CH_2CH_2I + NaBr\downarrow$$

This same compound can be made to undergo a bimolecular (E2) elimination reaction, another common type of second-order reaction, if the sodium iodide and acetone are replaced with sodium tert-butoxide as the salt and tert-butanol as the solvent:

$$CH_3CH_2CH_2CH_2Br + NaOt\text{-}Bu \rightarrow CH_3CH_2CH{=}CH_2 + NaBr + HOt\text{-}Bu$$

Pseudo-first Order

If the concentration of a reactant remains constant (because it is a catalyst, or because it is in great excess with respect to the other reactants), its concentration can be included in the rate constant, obtaining a *pseudo–first-order* (or occasionally pseudo–second-order) rate equation. For a typical second-order reaction with rate equation v = k[A][B], if the concentration of reactant B is constant then v = k[A][B] = k'[A], where the pseudo–first-order rate constant k' = k[B]. The second-order rate equation has been reduced to a pseudo–first-order rate equation, which makes the treatment to obtain an integrated rate equation much easier.

One way to obtain a pseudo-first order reaction is to use a large excess of one reactant (say, [B]≫[A]) so that, as the reaction progresses, only a small fraction of the reactant in excess (B) is consumed, and its concentration can be considered to stay constant.

For example, the hydrolysis of esters by dilute mineral acids follows pseudo-first order kinetics where the concentration of water is present in large excess:

$$CH_3COOCH_3 + H_2O \rightarrow CH_3COOH + CH_3OH$$

The hydrolysis of sucrose in acid solution is often cited as a first-order reaction with rate r = k[sucrose]. The true rate equation is third-order, r = k[sucrose][H^+][H_2O]; however, the concentrations of both the catalyst H^+ and the solvent H_2O are normally constant, so that the reaction is pseudo–first-order.

Reaction Orders 0, 1, 2 and n

Elementary reaction steps with order 3 (called *ternary reactions*) are rare and unlikely to occur. However, overall reactions composed of several elementary steps can, of course, be of any (including non-integer) order.

	Zero order	First order	Second order	nth order
Rate Law	$-d[A]/dt = k$	$-d[A]/dt = k[A]$	$-d[A]/dt = k[A]^2$	$-d[A]/dt = k[A]^n$
Integrated Rate Law	$[A] = [A]_0 - kt$	$[A] = [A]_0 e^{-kt}$	$\frac{1}{[A]} = \frac{1}{[A]_0} + kt$	$\frac{1}{[A]^{n-1}} = \frac{1}{[A]_0^{n-1}} + (n-1)kt$ [Except first order]
Units of Rate Constant (k)	$\frac{M}{s}$	$\frac{1}{s}$	$\frac{1}{M \cdot s}$	$\frac{1}{M^{n-1} \cdot s}$
Linear Plot to determine k	[A] vs. t	ln[A] vs. t	$\frac{1}{[A]}$ vs. t	$\frac{1}{[A]^{n-1}}$ vs. t [Except first order]
Half-life	$t_{\frac{1}{2}} = \frac{[A]_0}{2k}$	$t_{\frac{1}{2}} = \frac{\ln(2)}{k}$	$t_{\frac{1}{2}} = \frac{1}{k[A]_0}$	$t_{\frac{1}{2}} = \lim_{x \to n} \frac{2^{x-1} - 1}{(x-1)k[A]_0^{x-1}}$ [limit only necessary in first order]

Where M stands for concentration in molarity (mol · L^{-1}), *t* for time, and *k* for the reaction rate constant. The half-life of a first order reaction is often expressed as $t_{1/2} = 0.693/k$ (as ln2 = 0.693).

Fractional Order

In fractional order reactions, the order is a non-integer, which often indicates a chemical chain reaction or other complex reaction mechanism. For example, the pyrolysis of acetaldehyde (CH_3CHO) into methane and carbon monoxide proceeds with an order of 1.5 with respect to acetaldehyde: $r = k[CH_3CHO]^{3/2}$. The decomposition of phosgene ($COCl_2$) to carbon monoxide and chlorine has order 1 with respect to phosgene itself and order 0.5 with respect to chlorine: $v = k[COCl_2]\,[Cl_2]^{1/2}$.

The order of a chain reaction can be rationalized using the steady state approximation for the concentration of reactive intermediates such as free radicals. For the pyrolysis of acetaldehyde, the Rice-Herzfeld mechanism is:

Initiation:

$$CH_3CHO \rightarrow \bullet CH_3 + \bullet CHO$$

Propagation:

$$\bullet CH_3 + CH_3CHO \rightarrow CH_3CO\bullet + CH_4$$

$$CH_3CO\bullet \rightarrow \bullet CH_3 + CO$$

Termination:

$$2\ \bullet CH_3 \rightarrow C_2H_6$$

where • denotes a free radical. To simplify the theory, the reactions of the •CHO to form a second $\bullet CH_3$ are ignored.

In the steady state, the rates of formation and destruction of methyl radicals are equal, so that:

$$\frac{d[\cdot CH3]}{dt} = k_i[CH_3\,CHO] - k_t[\cdot CH_3]^2 = 0,$$

so that the concentration of methyl radical satisfies:

$$[\cdot CH_3] \ \propto \ [CH_3\,CHO]^{\frac{1}{2}}.$$

The reaction rate equals the rate of the propagation steps which form the main reaction products CH_4 and CO:

$$\upsilon = \frac{d[CH_4]}{dt} = k_p[\cdot CH_3][CH_3\,CHO] \ \propto \ [CH_3\,CHO]^{\frac{3}{2}}$$

in agreement with the experimental order of 3/2.

Complex Laws

Mixed Order

More complex rate laws have been described as being *mixed order* if they approximate to the laws for more than one order at different concentrations of the chemical species involved. For example, a rate law of the form $\upsilon = k_1[A] + k_2[A]^2$ represents concurrent first order and second order reactions (or more often concurrent pseudo-first order and second order) reactions, and can be described as mixed first and second order. For sufficiently large values of [A] such a reaction will approximate second order kinetics, but for smaller [A] the kinetics will approximate first order (or pseudo-first order). As the reaction progresses, the reaction can change from second order to first order as reactant is consumed.

Another type of mixed-order rate law has a denominator of two or more terms, often because the identity of the rate-determining step depends on the values of the concentrations. An example is the oxidation of an alcohol to a ketone by hexacyanoferrate (III) ion [$Fe(CN)_6^{3-}$] with ruthenate (VI) ion (RuO_4^{2-}) as catalyst. For this reaction, the rate of disappearance of hexacyanoferrate (III) is $v = \frac{[Fe(CN)_6]^{2-}}{k_\alpha + k_\beta [Fe(CN)_6]^{2-}}$.

This is zero-order with respect to hexacyanoferrate (III) at the onset of the reaction (when its concentration is high and the ruthenium catalyst is quickly regenerated), but changes to first-order when its concentration decreases and the regeneration of catalyst becomes rate-determining.

Notable mechanisms with mixed-order rate laws with two-term denominators include:

- Michaelis-Menten kinetics for enzyme-catalysis: First-order in substrate (second-order overall) at low substrate concentrations, zero order in substrate (first-order overall) at higher substrate concentrations.
- The Lindemann mechanism for unimolecular reactions: Second-order at low pressures, first-order at high pressures.

Negative Order

A reaction rate can have a negative partial order with respect to a substance. For example, the conversion of ozone (O_3) to oxygen follows the rate equation $v = k\frac{[O_3]^2}{[O_2]}$ in an excess of oxygen. This corresponds to second order in ozone and order (-1) with respect to oxygen.

When a partial order is negative, the overall order is usually considered as undefined. In the above example for instance, the reaction is not described as first order even though the sum of the partial orders is $2+(-1)=1$,, because the rate equation is more complex than that of a simple first-order reaction.

Opposed Reactions

A pair of forward and reverse reactions may occur simultaneously with comparable speeds. For example, A and B react into X and Y and vice versa (*s, t, u,* and *w* are the stoichiometric coefficients):

$$sA + tB \rightleftharpoons uX + wY$$

The reaction rate expression for the above reactions (assuming each one is elementary) can be expressed as:

$$v = k_1[\mathrm{A}]^s[\mathrm{B}]^t - k_2[\mathrm{X}]^u[\mathrm{Y}]^w$$

where: k_1 is the rate coefficient for the reaction that consumes A and B; k_2 is the rate coefficient for the backwards reaction, which consumes X and Y and produces A and B.

The constants k_1 and k_2 are related to the equilibrium coefficient for the reaction (K) by the following relationship (set v=0 in balance):

$$k_1[\mathrm{A}]^s[\mathrm{B}]^t = k_2[\mathrm{X}]^u[\mathrm{Y}]^w$$

$$K = \frac{[\mathrm{X}]^u[\mathrm{Y}]^w}{[\mathrm{A}]^s[\mathrm{B}]^t} = \frac{k_1}{k_2}$$

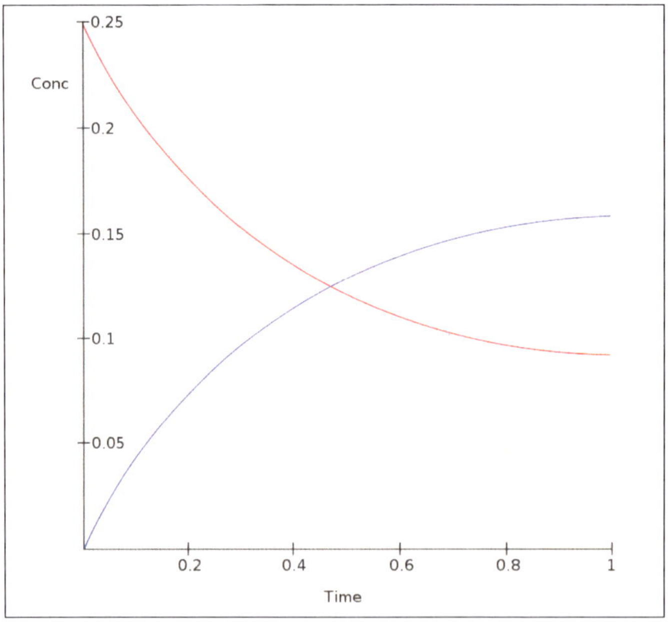

Concentration of A (A_0 = 0.25 mole/l) and B versus time reaching equilibrium k_f = 2 min^{-1} and k_r = 1 min^{-1}.

Simple Example

In a simple equilibrium between two species:

$$A \rightleftharpoons B$$

Where the reactions starts with an initial concentration of A, $[A]_0$, with an initial concentration of 0 for B at time t=0.

Then the constant K at equilibrium is expressed as:

$$K \stackrel{def}{=} \frac{k_f}{k_b} = \frac{[B]_e}{[A]_e}$$

Where $[A]_e$ and $[B]_e$ are the concentrations of A and B at equilibrium, respectively.

The concentration of A at time t, $[A]_t$, is related to the concentration of B at time t, $[B]_t$, by the equilibrium reaction equation:

$$[A]_t = [A]_0 - [B]_t$$

The term $[B]_0$ is not present because, in this simple example, the initial concentration of B is 0.

This applies even when time t is at infinity; i.e., equilibrium has been reached:

$$[A]_e = [A]_0 - [B]_e$$

then it follows, by the definition of K, that:

$$[B]_e = x = \frac{k_f}{k_f + k_b}[A]_0$$

and, therefore,

$$[A]_e = [A]_0 - x = \frac{k_b}{k_f + k_b}[A]_0$$

These equations allow us to uncouple the system of differential equations, and allow us to solve for the concentration of A alone.

The reaction equation, given previously as:

$$v = k_1[A]^s[B]^t - k_2[X]^u[Y]^v$$

$$-\frac{d[A]}{dt} = k_f[A]_t - k_b[B]_t$$

The derivative is negative because this is the rate of the reaction going from A to B, and therefore the concentration of A is decreasing. To simplify annotation, let x be

$[A]_t$, the concentration of A at time t. Let x_e be the concentration of A at equilibrium. Then:

$$d-\frac{d[A]}{dt}=k_f[A]_t-k_b[B]_t$$

$$-\frac{dx}{dt}=k_f x-k_b[B]_t$$

$$=k_f x-k_b([A]_0-x)$$

$$=(k_f+k_b)x-k_b[A]_0$$

Since:

$$k_f+k_b=k_b\frac{[A]_0}{x_e}$$

The reaction rate becomes:

$$\frac{dx}{dt}=\frac{k_b[A]_0}{x_e}(x_e-x)$$

which results in:

$$\ln\left(\frac{[A]_0-[A]_e}{[A]_t-[A]_e}\right)=(k_f+k_b)t$$

A plot of the negative natural logarithm of the concentration of A in time minus the concentration at equilibrium versus time t gives a straight line with slope $k_f + k_b$. By measurement of A_e and B_e the values of K and the two reaction rate constants will be known.

Generalization of Simple Example

If the concentration at the time $t = 0$ is different from above, the simplifications above are invalid, and a system of differential equations must be solved. However, this system can also be solved exactly to yield the following generalized expressions:

$$[A]=[A]_0\frac{1}{k_f+k_b}\left(k_b+k_f e^{-(k_f+k_b)t}\right)+[B]_0\frac{k_b}{k_f+k_b}\left(1-e^{-(k_f+k_b)t}\right)$$

$$[B]=[A]_0\frac{k_f}{k_f+k_b}\left(1-e^{-(k_f+k_b)t}\right)+[B]_0\frac{1}{k_f+k_b}\left(k_f+k_b e^{-(k_f+k_b)t}\right)$$

When the equilibrium constant is close to unity and the reaction rates very fast for instance in conformational analysis of molecules, other methods are required for the determination of rate constants for instance by complete lineshape analysis in NMR spectroscopy.

Consecutive Reactions

If the rate constants for the following reaction are k_1 and k_2 ; $A \rightarrow B \rightarrow C$,, then the rate equation is:

For reactant A: $\frac{d[A]}{dt} = -k_1[A]$

For reactant B: $\frac{d[B]}{dt} = k_1[A] - k_2[B]$

For product C: $\frac{d[C]}{dt} = k_2[B]$

With the individual concentrations scaled by the total population of reactants to become probabilities, linear systems of differential equations such as these can be formulated as a master equation. The differential equations can be solved analytically and the integrated rate equations are:

$$[A] = [A]_0 e^{-k_1 t}$$

$$[B] = \begin{cases} [A]_0 \frac{k_1}{k_2 - k_1}\left(e^{-k_1 t} - e^{-k_2 t}\right) + [B]_0 e^{-k_2 t} & k_1 \neq k_2 \\ [A]_0 k_1 t e^{-k_1 t} + [B]_0 e^{-k_1 t} \end{cases}$$

$$[C] = \begin{cases} [A]_0 \left(1 + \frac{k_1 e^{-k_2 t} - k_2 e^{-k_1 t}}{k_2 - k_1}\right) + [B]_0 \left(1 - e^{-k_2 t}\right) + [C]_0 & k_1 \neq k_2 \\ [A]_0 \left(1 - e^{-k_1 t} - k_1 t e^{-k_1 t}\right) + [B]_0 \left(1 - e^{-k_1 t}\right) + [C]_0 \end{cases}$$

The steady state approximation leads to very similar results in an easier way.

Parallel or Competitive Reactions

When a substance reacts simultaneously to give two different products, a parallel or competitive reaction is said to take place.

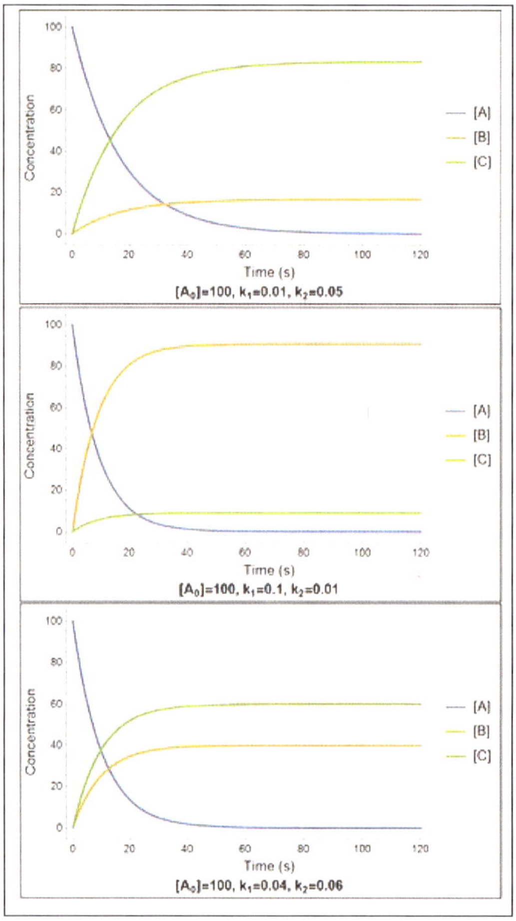

Time course of two first order, competitive reactions with differing rate constants.

Two First Order Reactions

$\mathrm{A} \rightarrow \mathrm{B}$ and $\mathrm{A} \rightarrow \mathrm{C}$, , with constants k_1 and K_2 and rate equations:

$$-\frac{d[\mathrm{A}]}{dt} = (k_1 + k_2)[\mathrm{A}]\,;\ \frac{d[\mathrm{B}]}{dt} = k_1[\mathrm{A}] \text{ and } \frac{d[\mathrm{C}]}{dt} = k_2[\mathrm{A}]$$

The integrated rate equations are then:

$$[\mathrm{A}] = [\mathrm{A}]_0 e^{-(k_1+k_2)t};$$

$$[\mathrm{B}] = \frac{k_1}{k_1+k_2}[\mathrm{A}]_0(1-e^{-(k_1+k_2)t})$$

and $$[\mathrm{C}] = \frac{k_2}{k_1+k_2}[\mathrm{A}]_0(1-e^{-(k_1+k_2)t}).$$

One important relationship in this case is $\frac{[\mathrm{B}]}{[\mathrm{C}]} = \frac{k_1}{k_2}$

One First Order and One Second Order Reaction

This can be the case when studying a bimolecular reaction and a simultaneous hydrolysis (which can be treated as pseudo order one) takes place: the hydrolysis complicates the study of the reaction kinetics, because some reactant is being "spent" in a parallel reaction. For example, A reacts with R to give our product C, but meanwhile the hydrolysis reaction takes away an amount of A to give B, a byproduct: $\mathrm{A+H_2O \rightarrow B}$ and $\mathrm{A+R \rightarrow C}$. The rate equations are: $\frac{d[\mathrm{B}]}{dt} = k_1[\mathrm{A}][\mathrm{H_2O}] = k_1^{'}[\mathrm{A}]$ and $\frac{d[\mathrm{C}]}{dt} = k_2[\mathrm{A}][\mathrm{R}]$.
Where $k_1^{'}$ is the pseudo first order constant.

The integrated rate equation for the main product [C] is $[\mathrm{C}] = [\mathrm{R}]_0 \left[1 - e^{-\frac{k_2}{k'_1}[\mathrm{A}]_0(1-e^{-k_1' t})} \right]$,
which is equivalent to $\ln \frac{[\mathrm{R}]_0}{[\mathrm{R}]_0 - [\mathrm{C}]} = \frac{k_2[\mathrm{A}]_0}{k_1^{'}}(1-e^{-k_1^{'}t})$. Concentration of B is related to that of C through $[\mathrm{B}] = -\frac{k_1^{'}}{k_2} ln\left(1 - \frac{[\mathrm{C}]}{[\mathrm{R}]_0}\right)$.

The integrated equations were analytically obtained but during the process it was assumed that $[\mathrm{A}]_0 - [\mathrm{C}] \approx [\mathrm{A}]_0$ therefeore, previous equation for [C] can only be used for low concentrations of [C] compared to $[\mathrm{A}]_0$

Stoichiometric Reaction Networks

The most general description of a chemical reaction network considers a number N of distinct chemical species reacting via R reactions. The chemical equation of the j-th reaction can then be written in the generic form:

$$s_{1j}\,\mathrm{X}_1 + s_{2j}\,\mathrm{X}_2 \ldots + s_{Nj}\,\mathrm{X}_N \xrightarrow{k_j} r_{1j}\,\mathrm{X}_1 + r_{2j}\,\mathrm{X}_2 + \ldots + r_{Nj}\,\mathrm{X}_N,$$

which is often written in the equivalent form:

$$\sum_{i=1}^{N} s_{ij}\,\mathrm{X}_i \xrightarrow{k_j} \sum_{i=1}^{N} r_{ij}\,\mathrm{X}_i.$$

Here,

- j is the reaction index running from 1 to R,
- X_i denotes the j – th chemical species,
- k_j is the rate constant of the j-th reaction,
- s_{ij} and r_{ij} are the stoichiometric coefficients of reactants and products, respectively.

The rate of such reaction can be inferred by the law of mass action:

$$f_j([\vec{X}]) = k_j \prod_{z=1}^{N} [X_z]^{s_{zj}}$$

which denotes the flux of molecules per unit time and unit volume. Here $[\vec{X}] = ([X_1],[X_2],...,[X_N])$ is the vector of concentrations. This definition includes the elementary reactions:

- Zero Order Reactions:

  for which $s_{zj} = 0$ for all z,

- First Order Reactions:

  for which $s_{zj} = 1$ for a single z,

- Second Order Reactions:

  for which $s_{zj} = 1$ for exactly two z, i.e., a bimolecular reaction, or $s_{zj} = 2$ for a single z i.e., a dimerization reaction.

One can define the stoichiometric matrix:

$$S_{ij} = r_{ij} - s_{ij},$$

denoting the net extent of molecules of i in reaction j. The reaction rate equations can then be written in the general form:

$$\frac{d[X_i]}{dt} = \sum_{j=1}^{R} S_{ij} f_j([\vec{X}]).$$

This is the product of the stoichiometric matrix and the vector of reaction rate functions. Particular simple solutions exist in equilibrium, $\frac{d[X_i]}{dt} = 0$, for systems composed of merely reversible reactions. In this case the rate of the forward and backward reactions are equal, a principle called detailed balance. Detailed balance is a property of the stoichiometric matrix S_{ij} alone and does not depend on the particular form of the rate functions f_j. All other cases where detailed balance is violated are commonly studied by flux balance analysis which has been developed to understand metabolic pathways.

Dynamics of Unimolecular Conversion

For a general unimolecular reaction involving interconversion of N different species, whose concentrations at time t are denoted by $X_1(t)$ through $X_N(t)$, an analytic form for the time-evolution of the species can be found. Let the rate constant of conversion from species X_i to species X_j be denoted as K_{ij}, and construct a rate-constant matrix K whose entries are the K_{ij}.

Also, let $X(t) = (X_1(t), X_2(t), ..., X_N(t))^T$ be the vector of concentrations as a function of time.

Let $J = (1,1,1,...,1)^T$ be the vector of ones.

Let I be the $N \times N$ identity matrix.

Let Diag be the function that takes a vector and constructs a diagonal matrix whose on-diagonal entries are those of the vector.

Let $\mathcal{L}^{-1}$ be the inverse Laplace transform from s to t.

Then the time-evolved state $X(t)$ is given by:

$$X(t) = \mathcal{L}^{-1}[(sI + Diag(KJ) - K^T)^{-1} X(0)],$$

thus providing the relation between the initial conditions of the system and its state at time t.

Rate-determining Step

In chemical kinetics, the overall rate of a reaction is often approximately determined by the slowest step, known as the rate-determining step (RDS) or rate-limiting step. For a given reaction mechanism, the prediction of the corresponding rate equation (for comparison with the experimental rate law) is often simplified by using this approximation of the rate-determining step.

In principle, the time evolution of the reactant and product concentrations can be determined from the set of simultaneous rate equations for the individual steps of the mechanism, one for each step. However, the analytical solution of these differential equations is not always easy, and in some cases numerical integration may even be required. The hypothesis of a single rate-determining step can greatly simplify the mathematics. In the simplest case the initial step is the slowest, and the overall rate is just the rate of the first step.

Also, the rate equations for mechanisms with a single rate-determining step are usually in a simple mathematical form, whose relation to the mechanism and choice of rate-determining step is clear. The correct rate-determining step can be identified by predicting the rate law for each possible choice and comparing the different predictions with the experimental law, as for the example of NO_2 and CO below.

The concept of the rate-determining step is very important to the optimization and understanding of many chemical processes such as catalysis and combustion.

Example Reaction: NO_2 + CO

As an example, consider the gas-phase reaction $NO_2 + CO \rightarrow NO + CO_2$. If this reaction occurred in a single step, its reaction rate (r) would be proportional to the rate of collisions between NO_2 and CO molecules: $r = k[NO_2][CO]$, where k is the reaction rate constant, and square brackets indicate a molar concentration. Another typical example is the Zel'dovich mechanism.

First Step Rate-determining

In fact, however, the observed reaction rate is second-order in NO_2 and zero-order in CO,with rate equation $r = k[NO_2]^2$. This suggests that the rate is determined by a step in which two NO_2 molecules react, with the CO molecule entering at another, faster, step. A possible mechanism in two elementary steps that explains the rate equation is:

$NO_2 + NO_2 \rightarrow NO + NO_3$ (slow step, rate-determining).

$NO_3 + CO \rightarrow NO_2 + CO_2$ (fast step).

In this mechanism the reactive intermediate species NO_3 is formed in the first step with rate r_1 and reacts with CO in the second step with rate r_2. However NO_3 can also react with NO if the first step occurs in the *reverse direction* ($NO + NO_3 \rightarrow 2\ NO_2$) with rate r_{-1}, where the minus sign indicates the rate of a reverse reaction.

The concentration of a reactive intermediate such as $[NO_3]$ remains low and almost constant. It may therefore be estimated by the steady-state approximation, which specifies that the rate at which it is formed equals the (total) rate at which it is consumed. In this example NO_3 is formed in one step and reacts in two, so that:

$$\frac{d[NO_3]}{dt} = r_1 - r_2 - r_{-1} \approx 0.$$

The statement that the first step is the slow step actually means that the first step *in the reverse direction* is slower than the second step in the forward direction, so that almost all NO_3 is consumed by reaction with CO and not with NO. That is, $r_{-1} \ll r_2$, so that $r_1 - r_2 \approx 0$. But the overall rate of reaction is the rate of formation of final product (here CO_2), so that $r = r_2 \approx r_1$. That is, the overall rate is determined by the rate of the first step, and (almost) all molecules that react at the first step continue to the fast second step.

Pre-equilibrium: If the Second Step were Rate-determining

The other possible case would be that the second step is slow and rate-determining, meaning that it is slower than the first step in the reverse direction: $r_2 \ll r_{-1}$. In this hypothesis, $r_1 - r_{-1} \approx 0$, so that the first step is (almost) at equilibrium. The overall rate is determined by the second step: $r = r_2 \ll r_1$, as very few molecules that react at the first step continue to the second step, which is much slower. Such a situation in which an intermediate (here NO_3) forms an equilibrium with reactants *prior* to the rate-determining step is described as a *pre-quilibrium*. For the reaction of NO_2 and CO, this hypothesis can be rejected, since it implies a rate equation that disagrees with experiment.

If the first step were at equilibrium, then its equilibrium constant expression permits calculation of the concentration of the intermediate NO_3 in terms of more stable (and more easily measured) reactant and product species:

$$K_1 = \frac{[NO][NO_3]}{[NO_2]^2},$$

$$[NO_3] = K_1 \frac{[NO_2]^2}{[NO]}.$$

The overall reaction rate would then be:

$$r = r_2 = k_2[NO_3][CO] = k_2 K_1 \frac{[NO_2]^2[CO]}{[NO]},$$

which disagrees with the experimental rate law given above, and so disproves the hypothesis that the second step is rate-determining for this reaction. However, some other reactions are believed to involve rapid pre-equilibria prior to the rate-determining step.

Nucleophilic Substitution

Another example is the unimolecular nucleophilic substitution (S_N1) reaction in organic chemistry, where it is the first, rate-determining step that is unimolecular. A specific case is the basic hydrolysis of tert-butyl bromide (t-C_4H_9Br) by aqueous sodium hydroxide. The mechanism has two steps (where R denotes the tert-butyl radical t-C_4H_9):

1. Formation of a carbocation $R{-}Br \rightarrow R^+ + Br^-$.
2. Nucleophilic attack by one water molecule $R^+ + OH^- \rightarrow ROH$.

This reaction is found to be first-order with $r = k[R{-}Br]$, which indicates that the first step is slow and determines the rate. The second step with OH^- is much faster, so the overall rate is independent of the concentration of OH^-.

In contrast, the alkaline hydrolysis of methyl bromide (CH_3Br) is a bimolecular nucleophilic substitution (S_N2) reaction in a single bimolecular step. Its rate law is second-order: $r = k[R{-}Br][OH^-]$.

Composition of the Transition State

A useful rule in the determination of mechanism is that the concentration factors in the rate law indicate the composition and charge of the activated complex or transition state. For the NO_2 –CO reaction above, the rate depends on $[NO_2]^2$, so that the activated complex has composition N_2O_4, with 2 NO_2 entering the reaction before the transition state, and CO reacting after the transition state.

A multistep example is the reaction between oxalic acid and chlorine in aqueous solution: $H_2C_2O_4 + Cl_2 \rightarrow 2\,CO_2 + 2\,H^+ + 2\,Cl^-$. The observed rate law is:

$$v = k\frac{[Cl_2][H_2C_2O_4]}{[H^+]^2[Cl^-]},$$

which implies an activated complex in which the reactants lose $2H^+ + Cl^-$ before the rate-determining step. The formula of the activated complex is:

$$Cl_2 + H_2C_2O_4 - 2\,H^+ - Cl^- + X\,H_2O, \text{or } C_2O_4Cl(H_2O)_x^-$$

(an unknown number of water molecules are added because the possible dependence of the reaction rate on H_2O was not studied, since the data were obtained in water solvent at a large and essentially unvarying concentration).

One possible mechanism in which the preliminary steps are assumed to be rapid pre-equilibria occurring prior to the transition state is:

$$Cl_2 + H_2O \rightleftharpoons HOCl + Cl^- + H^+$$

$$H_2C_2O_4 \rightleftharpoons H^+ + HC_2O^-_4$$

$$HOCl + HC_2O_4^- \rightarrow H_2O + Cl^- + 2\,CO_2$$

Reaction Coordinate Diagram

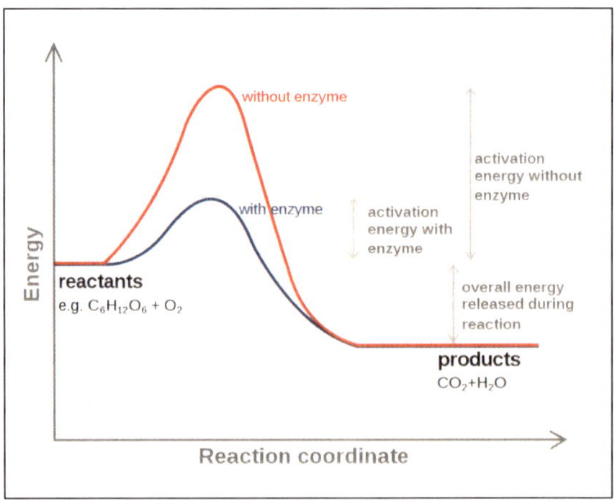

In a multistep reaction, the rate-determining step does not necessarily correspond to the highest Gibbs energy on the reaction coordinate diagram. If there is a reaction intermediate whose energy is lower than the initial reactants, then the activation energy needed to pass through any subsequent transition state depends on the Gibbs energy of that state relative to the lower-energy intermediate. The rate-determining step is then the step with the largest Gibbs energy difference relative either to the starting material or to any previous intermediate on the diagram.

Also, for reaction steps that are not first-order, concentration terms must be considered in choosing the rate-determining step.

Chain Reactions

Not all reactions have a single rate-determining step. In particular, the rate of a chain reaction is usually not controlled by any single step.

Diffusion Control

In the previous examples the rate determining step was one of the sequential chemical reactions leading to a product. The rate-determining step can also be the transport of reactants to where they can interact and form the product. This case is referred to as diffusion control and, in general, occurs when the formation of product from the activated complex is very rapid and thus the provision of the supply of reactants is rate-determining.

Law of Mass Action

The law of mass action states that the rate of a reaction is proportional to the product of the concentrations of each reactant. This law can be used to explain the behaviour exhibited by solutions in dynamic equilibria. The law of mass action also suggests that the ratio of the reactant concentration and the product concentration is constant at a state of chemical equilibrium.

The Equilibrium Constant (K_c)

The concentration of reactants and products, at equilibrium, are constant at a given temperature. Consider the following simple reversible reaction where A & B are the reactants whereas C & D are the products.

$$A + B \rightleftharpoons C + D$$

A mixture of products and reactants in a state of chemical equilibrium is known as an equilibrium mixture. There exists a relation between the concentration of products and the concentration of reactants for an equilibrium mixture. This relation can be equated as follows.

$$K_c = \frac{[C][D]}{[A][B]}$$

Here, K_c is called the equilibrium constant. In this equation, the concentration of A at equilibrium is represented as [A] (similarly for B, C, and D), and the stoichiometric coefficients of the reactants and products are 1. It has been experimentally observed that the equilibrium constant is also dependant on the stoichiometric coefficients of the reactants and products.

Therefore, the law of mass action dictates that the equilibrium constant, at a given constant temperature, is equal to the product of the concentration of products raised to the respective stoichiometric coefficients divided by the product of the reactant concentrations, each raised to the corresponding stoichiometric coefficient.

This is also known as the equilibrium law or the law of chemical equilibrium. Representation of the Equilibrium Constant. For a balanced reaction of the type,

$$aA + bB \rightleftharpoons cC + dD$$

According to the law of mass action, the constant value obtained by relating equilibrium concentrations of reactants and products is called the equilibrium constant. For the forward reaction, this is given by:

$$K_c = \frac{[C]^c[D]^d}{[A]^a[B]^b}$$

The equilibrium constant for the reverse reaction is the inverse of the forward reaction and is given by:

$$K_c' = \frac{1}{K_c} = \frac{[A]^a[B]^b}{[C]^c[D]^d}$$

If the coefficients of the chemical equation are multiplied by a factor 'n' then the equilibrium constant is raised by the power 'n' i.e. the constant becomes K_c^n.

Equilibrium Constant Representation	Expressed in terms of	Expressed as
K_c	Concentrations of reactants and products.	$\frac{[C]^c[D]^d}{[A]^a[B]^b}$
K_p	Partial pressures of reactants and products. (only for the substances which are in gaseous state).	$\frac{p_C^c p_D^d}{p_A^a p_B^b}$

K_x	Mole fractions of reactants and products.	$\frac{[X_C]^c[X_D]^d}{[X_A]^a[X_B]^b}$

Relation between K_c, K_p and K_x:

$$K_p = K_c(RT)^{\Delta n_g}$$

$$K_x = K_p(RT)^{\Delta n_g}$$

where, Δn_g = moles of gaseous products – moles of gaseous reactants. K_c is the equilibrium constant expressed in terms of the concentration of the reactants/products. Similarly, K_p is the constant in terms of the partial pressures of the substances and K_x is expressed in terms of the mole fraction.

Chemical Equilibrium

In a chemical reaction, chemical equilibrium is the state in which both reactants and products are present in concentrations which have no further tendency to change with time, so that there is no observable change in the properties of the system. Usually, this state results when the forward reaction proceeds at the same rate as the reverse reaction. The reaction rates of the forward and backward reactions are generally not zero, but equal. Thus, there are no net changes in the concentrations of the reactant(s) and product(s). Such a state is known as dynamic equilibrium.

Thermodynamics

At constant temperature and pressure, one must consider the Gibbs free energy, *G*, while at constant temperature and volume, one must consider the Helmholtz free energy, *A*, for the reaction; and at constant internal energy and volume, one must consider the entropy, *S*, for the reaction.

The constant volume case is important in geochemistry and atmospheric chemistry where pressure variations are significant. Note that, if reactants and products were in standard state (completely pure), then there would be no reversibility and no equilibrium. Indeed, they would necessarily occupy disjoint volumes of space. The mixing of the products and reactants contributes a large entropy (known as entropy of mixing) to states containing equal mixture of products and reactants. The standard Gibbs energy change, together with the Gibbs energy of mixing, determine the equilibrium state.

Here only the constant pressure case is considered. The relation between the Gibbs free energy and the equilibrium constant can be found by considering chemical potentials.

At constant temperature and pressure, the Gibbs free energy, G, for the reaction depends only on the extent of reaction: ξ, and can only decrease according to the second law of thermodynamics. It means that the derivative of G with ξ must be negative if the reaction happens; at the equilibrium the derivative being equal to zero.

$$\left(\frac{dG}{d\xi}\right)_{T,p} = 0 \text{ : equilibrium}$$

In order to meet the thermodynamic condition for equilibrium, the Gibbs energy must be stationary, meaning that the derivative of G with respect to the extent of reaction: ξ, must be zero. It can be shown that in this case, the sum of chemical potentials of the products is equal to the sum of those corresponding to the reactants. Therefore, the sum of the Gibbs energies of the reactants must be the equal to the sum of the Gibbs energies of the products.

$$\alpha\mu_{\mathrm{A}} + \beta\mu_{\mathrm{B}} = \sigma\mu_{\mathrm{S}} + \tau\mu_{\mathrm{T}}$$

where μ is in this case a partial molar Gibbs energy, a chemical potential. The chemical potential of a reagent A is a function of the activity, {A} of that reagent.

$$\mu_{\mathrm{A}} = \mu_A^{\ominus} + RT\ln\{\mathrm{A}\}$$

(where $\mu_A^{\ominus}$ is the standard chemical potential).

The definition of the Gibbs energy equation interacts with the fundamental thermodynamic relation to produce:

$$dG = Vdp - SdT + \sum_{i=1}^{k} \mu_i dN_i .$$

Inserting $dN_i = v_i\, d\xi$ into the above equation gives a Stoichiometric coefficient (υ_i) and a differential that denotes the reaction occurring once ($d\xi$). At constant pressure and temperature the above equations can be written as:

$$\left(\frac{dG}{d\xi}\right)_{T,p} = \sum_{i=1}^{k} \mu_i v_i = \Delta_r G_{T,p}$$

which is the "Gibbs free energy change for the reaction.

This results in:

$$\Delta_r G_{T,p} = \sigma\mu_S + \tau\mu_{\mathrm{T}} - \alpha\mu_{\mathrm{A}} - \beta\mu_{\mathrm{B}} .$$

By substituting the chemical potentials:

$$\Delta_r G_{T,p} = (\sigma\mu_{\mathrm{S}}^{\ominus} + \tau\mu_{\mathrm{T}}^{\ominus}) - (\alpha\mu_{\mathrm{A}}^{\ominus} + \beta\mu_{\mathrm{B}}^{\ominus}) + (\sigma RT\ln\{\mathrm{S}\} + \tau RT\ln\{\mathrm{T}\}) - (\alpha RT\ln\{\mathrm{A}\} + \beta RT\ln\{\mathrm{B}\}),$$

the relationship becomes:

$$\Delta_r G_{T,p} = \sum_{i=1}^{k} \mu_i^{\ominus} \nu_i + RT \ln \frac{\{S\}^{\sigma}\{T\}^{\tau}}{\{A\}^{\alpha}\{B\}^{\beta}}$$

$$\sum_{i=1}^{k} \mu_i^{\ominus} \nu_i = \Delta_r G^{\ominus}:$$

which is the standard Gibbs energy change for the reaction that can be calculated using thermodynamical tables. The reaction quotient is defined as:

$$Q_r = \frac{\{S\}^{\sigma}\{T\}^{\tau}}{\{A\}^{\alpha}\{B\}^{\beta}}$$

Therefore,

$$\left(\frac{dG}{d\xi}\right)_{T,p} = \Delta_r G_{T,p} = \Delta_r G^{\ominus} + RT \ln Q_r$$

At equilibrium:

$$\left(\frac{dG}{d\xi}\right)_{T,p} = \Delta_r G_{T,p} = 0$$

leading to:

$$0 = \Delta_r G^{\ominus} + RT \ln K_{eq}$$

and

$$\Delta_r G^{\ominus} = -RT \ln K_{eq}$$

Obtaining the value of the standard Gibbs energy change, allows the calculation of the equilibrium constant.

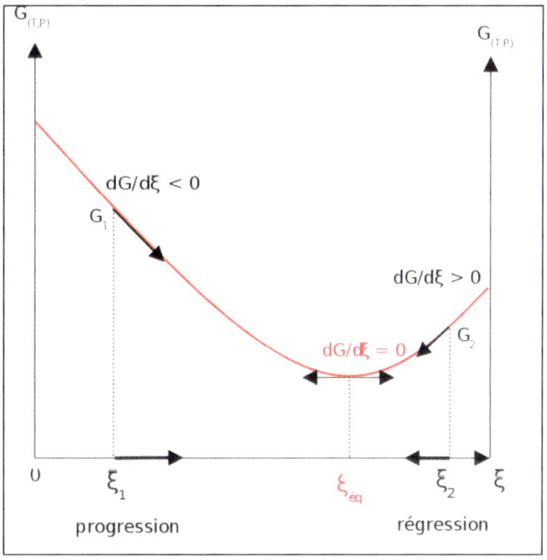

Addition of Reactants or Products

For a reactional system at equilibrium: $Q_r = K_{eq}$; $\xi = \xi_{eq}$.

- If are modified activities of constituents, the value of the reaction quotient changes and becomes different from the equilibrium constant: $Q_r \neq K_{eq}$:

  $$\left(\frac{dG}{d\xi}\right)_{T,p} = \Delta_r G^{\ominus} + RT \ln Q_r$$

  and

  $$\Delta_r G^{\ominus} = -RT \ln K_{eq}$$

  then:

  $$\left(\frac{dG}{d\xi}\right)_{T,p} = RT \ln\left(\frac{Q_r}{K_{eq}}\right)$$

- If activity of a reagent *i* increases:

  $$Q_r = \frac{\prod (a_j)^{\nu_j}}{\prod (a_i)^{\nu_i}}$$, the reaction quotient decreases.

  then:

  $$Q_r < K_{eq}$$

  and

  $$\left(\frac{dG}{d\xi}\right)_{T,p} < 0$$

  The reaction will shift to the right (i.e. in the forward direction, and thus more products will form).

- If activity of a product *j* increases:

  then:

  $$Q_r < K_{eq} \text{ and } \left(\frac{dG}{d\xi}\right)_{T,p} < 0$$

  The reaction will shift to the left (i.e. in the reverse direction, and thus less products will form).

Note that activities and equilibrium constants are dimensionless numbers.

Treatment of Activity

The expression for the equilibrium constant can be rewritten as the product of a concentration quotient, K_c and an activity coefficient quotient, Γ.

$$K = \frac{[\mathrm{S}]^{\sigma}[\mathrm{T}]^{\tau}\ldots}{[\mathrm{A}]^{\alpha}[\mathrm{B}]^{\beta}\ldots} \times \frac{\gamma_{\mathrm{S}}{}^{\sigma}\gamma_{\mathrm{T}}{}^{\tau}\ldots}{\gamma_{\mathrm{A}}{}^{\alpha}\gamma_{\mathrm{B}}{}^{\beta}\ldots} = K_c\Gamma$$

[A] is the concentration of reagent A, etc. It is possible in principle to obtain values of the activity coefficients, γ. For solutions, equations such as the Debye–Hückel equation or extensions such as Davies equation Specific ion interaction theory or Pitzer equations may be used. However this is not always possible. It is common practice to assume that Γ is a constant, and to use the concentration quotient in place of the thermodynamic equilibrium constant. It is also general practice to use the term *equilibrium constant* instead of the more accurate *concentration quotient*. This practice will be followed here.

For reactions in the gas phase partial pressure is used in place of concentration and fugacity coefficient in place of activity coefficient. In the real world, for example, when making ammonia in industry, fugacity coefficients must be taken into account. Fugacity, *f*, is the product of partial pressure and fugacity coefficient. The chemical potential of a species in the gas phase is given by:

$$\mu = \mu^{\ominus} + RT\ln\left(\frac{f}{bar}\right) = \mu^{\ominus} + RT\ln\left(\frac{p}{bar}\right) + RT\ln\gamma$$

so the general expression defining an equilibrium constant is valid for both solution and gas phases.

Concentration Quotients

In aqueous solution, equilibrium constants are usually determined in the presence of an "inert" electrolyte such as sodium nitrate $NaNO_3$ or potassium perchlorate $KClO_4$. The ionic strength of a solution is given by:

$$I = \frac{1}{2}\sum_{i=1}^{N} c_i z_i^2$$

where c_i and z_i stand for the concentration and ionic charge of ion type *i*, and the sum is taken over all the *N* types of charged species in solution. When the concentration of dissolved salt is much higher than the analytical concentrations of the reagents, the ions originating from the dissolved salt determine the ionic strength, and the ionic strength is effectively constant. Since activity coefficients depend on ionic strength the activity coefficients of the species are effectively independent of concentration. Thus, the assumption that Γ is constant is justified. The concentration quotient is a simple multiple of the equilibrium constant.

$$K_c = \frac{K}{\Gamma}$$

However, K_c will vary with ionic strength. If it is measured at a series of different ionic strengths the value can be extrapolated to zero ionic strength. The concentration quotient obtained in this manner is known, paradoxically, as a thermodynamic equilibrium constant.

To use a published value of an equilibrium constant in conditions of ionic strength different from the conditions used in its determination, the value should be adjusted.

Metastable Mixtures

A mixture may appear to have no tendency to change, though it is not at equilibrium. For example, a mixture of SO_2 and O_2 is metastable as there is a kinetic barrier to formation of the product, SO_3.

$$2\ SO_2 + O_2 \rightleftharpoons 2\ SO_3$$

The barrier can be overcome when a catalyst is also present in the mixture as in the contact process, but the catalyst does not affect the equilibrium concentrations.

Likewise, the formation of bicarbonate from carbon dioxide and water is very slow under normal conditions:

$$CO_2 + 2\ H_2O \rightleftharpoons HCO_3^- + H_3O^+$$

but almost instantaneous in the presence of the catalytic enzyme carbonic anhydrase.

Pure Substances

When pure substances (liquids or solids) are involved in equilibria their activities do not appear in the equilibrium constant because their numerical values are considered one.

Applying the general formula for an equilibrium constant to the specific case of a dilute solution of acetic acid in water one obtains:

$$CH_3CO_2H + H_2O \rightleftharpoons CH_3CO_2^- + H_3O^+$$

$$K_c = \frac{[CH_3CO_2^-][H_3O^+]}{[CH_3CO_2H][H_2O]}$$

For all but very concentrated solutions, the water can be considered a "pure" liquid, and therefore it has an activity of one. The equilibrium constant expression is therefore usually written as:

$$K = \frac{[CH_3CO_2^-][H_3O^+]}{[CH_3CO_2H]} = K_c.$$

A particular case is the self-ionization of water itself:

$$2\,H_2O \rightleftharpoons H_3O^+ + OH^-$$

Because water is the solvent, and has an activity of one, the self-ionization constant of water is defined as:

$$K_w = [H^+][OH^-]$$

It is perfectly legitimate to write $[H^+]$ for the hydronium ion concentration, since the state of solvation of the proton is constant (in dilute solutions) and so does not affect the equilibrium concentrations. K_w varies with variation in ionic strength and/or temperature.

The concentrations of H^+ and OH^- are not independent quantities. Most commonly $[OH^-]$ is replaced by $K_w[H^+]^{-1}$ in equilibrium constant expressions which would otherwise include hydroxide ion.

Solids also do not appear in the equilibrium constant expression, if they are considered to be pure and thus their activities taken to be one. An example is the Boudouard reaction:

$$2\,CO \rightleftharpoons CO_2 + C$$

for which the equation (without solid carbon) is written as:

$$K_c = \frac{[CO_2]}{[CO]^2}$$

Multiple Equilibria

Consider the case of a dibasic acid H_2A. When dissolved in water, the mixture will contain H_2A, HA^- and A^{2-}. This equilibrium can be split into two steps in each of which one proton is liberated.

$$H_2A \rightleftharpoons HA + H^+ : \quad K_1 = \frac{[HA^-][H^+]}{[H_2A]}$$

$$HA^- \rightleftharpoons A^{2-} + H^+ : \quad K_2 = \frac{[A^{2-}][H^+]}{[HA^-]}$$

K_1 and K_2 are examples of *stepwise* equilibrium constants. The *overall* equilibrium constant, β_D, is product of the stepwise constants.

$$H_2A \rightleftharpoons A^{2-} + 2H^+ : \ \beta_D = \frac{[A^{2-}][H^+]^2}{[H_2A]} = K_1K_2$$

Note that these constants are dissociation constants because the products on the right hand side of the equilibrium expression are dissociation products. In many systems, it is preferable to use association constants.

$$A^{2-} + H^+ \rightleftharpoons HA^- : \quad \beta_1 = \frac{[HA^-]}{[A^{2-}][H^+]}$$

$$A^{2-} + 2H^+ \rightleftharpoons H_2A : \ \beta_2 = \frac{[H_2A]}{[A^{2-}][H^+]^2}$$

β_1 and β_2 are examples of association constants. Clearly $\beta_1 = \frac{1}{K_2}$ and $\beta_2 = \frac{1}{\beta_D}$; $\log \beta_1 = pK_2$ and $\log \beta_2 = pK_2 + pK_1$.

Effect of Temperature

The effect of changing temperature on an equilibrium constant is given by the van 't Hoff equation:

$$\frac{d\ln K}{dT} = \frac{\Delta H_m^{\ominus}}{RT^2}$$

Thus, for exothermic reactions (ΔH is negative), K decreases with an increase in temperature, but, for endothermic reactions, (ΔH is positive) K increases with an increase temperature. An alternative formulation is:

$$\frac{d\ln K}{d(T^{-1})} = -\frac{\Delta H_m^{\ominus}}{R}$$

At first sight this appears to offer a means of obtaining the standard molar enthalpy of the reaction by studying the variation of K with temperature. In practice, however, the method is unreliable because error propagation almost always gives very large errors on the values calculated in this way.

Effect of Electric and Magnetic Fields

The effect of electric field on equilibrium has been studied by Manfred Eigen among others.

Types of Equilibrium

Haber–Bosch Process

N_2 (g) ⇌ N_2 (adsorbed)

N_2 (adsorbed) ⇌ 2 N (adsorbed)

H_2 (g) ⇌ H_2 (adsorbed)

H_2 (adsorbed) ⇌ 2 H (adsorbed)

N (adsorbed) + 3 H(adsorbed) ⇌ NH_3 (adsorbed)

NH_3 (adsorbed) ⇌ NH_3 (g)

Equilibrium can be broadly classified as heterogeneous and homogeneous equilibrium. Homogeneous equilibrium consists of reactants and products belonging in the same phase whereas heterogeneous equilibrium comes into play for reactants and products in different phases.

- In the gas phase: Rocket engines.
- The industrial synthesis such as ammonia in the Haber–Bosch process (depicted right) takes place through a succession of equilibrium steps including adsorption processes.
- Atmospheric chemistry.
- Seawater and other natural waters: Chemical oceanography.
- Distribution between two phases.
    - log *D* distribution coefficient: Important for pharmaceuticals where lipophilicity is a significant property of a drug.
    - Liquid–liquid extraction, Ion exchange, Chromatography.
    - Solubility product.
    - Uptake and release of oxygen by hemoglobin in blood.
- Acid–base equilibria: Acid dissociation constant, hydrolysis, buffer solutions, indicators, acid–base homeostasis.
- Metal–ligand complexation: Sequestering agents, chelation therapy, MRI contrast reagents, Schlenk equilibrium.
- Adduct formation: Host–guest chemistry, supramolecular chemistry, molecular recognition, dinitrogen tetroxide.

- In certain oscillating reactions, the approach to equilibrium is not asymptotically but in the form of a damped oscillation.
- The related Nernst equation in electrochemistry gives the difference in electrode potential as a function of redox concentrations.
- When molecules on each side of the equilibrium are able to further react irreversibly in secondary reactions, the final product ratio is determined according to the Curtin–Hammett principle.

In these applications, terms such as stability constant, formation constant, binding constant, affinity constant, association/dissociation constant are used. In biochemistry, it is common to give units for binding constants, which serve to define the concentration units used when the constant's value was determined.

Composition of a Mixture

When the only equilibrium is that of the formation of a 1:1 adduct as the composition of a mixture, there are many ways that the composition of a mixture can be calculated.

There are three approaches to the general calculation of the composition of a mixture at equilibrium.

1. The most basic approach is to manipulate the various equilibrium constants until the desired concentrations are expressed in terms of measured equilibrium constants (equivalent to measuring chemical potentials) and initial conditions.
2. Minimize the Gibbs energy of the system.
3. Satisfy the equation of mass balance. The equations of mass balance are simply statements that demonstrate that the total concentration of each reactant must be constant by the law of conservation of mass.

Mass-balance Equations

In general, the calculations are rather complicated or complex. For instance, in the case of a dibasic acid, H_2A dissolved in water the two reactants can be specified as the conjugate base, A^{2-}, and the proton, H^+. The following equations of mass-balance could apply equally well to a base such as 1,2-diaminoethane, in which case the base itself is designated as the reactant A:

$$T_A = [A]+[HA]+[H_2A]$$
$$T_H = [H]+[HA]+2[H_2A]-[OH]$$

With T_A the total concentration of species A. Note that it is customary to omit the ionic charges when writing and using these equations.

When the equilibrium constants are known and the total concentrations are specified there are two equations in two unknown "free concentrations" [A] and [H]. This follows from the fact that $[HA] = \beta_1[A][H]$, $[H_2A] = \beta_2[A][H]^2$ and $[OH] = K_w[H]^{-1}$:

$$T_A = [A] + \beta_1[A][H] + \beta_2[A][H]^2$$
$$T_H = [H] + \beta_1[A][H] + 2\beta_2[A][H]^2 - K_w[H]^{-1}$$

so the concentrations of the "complexes" are calculated from the free concentrations and the equilibrium constants. General expressions applicable to all systems with two reagents, A and B would be;

$$T_A = [A] + \sum_i p_i \beta_i [A]^{p_i} [B]^{q_i}$$
$$T_B = [B] + \sum_i q_i \beta_i [A]^{p_i} [B]^{q_i}$$

It is easy to see how this can be extended to three or more reagents.

Polybasic Acids

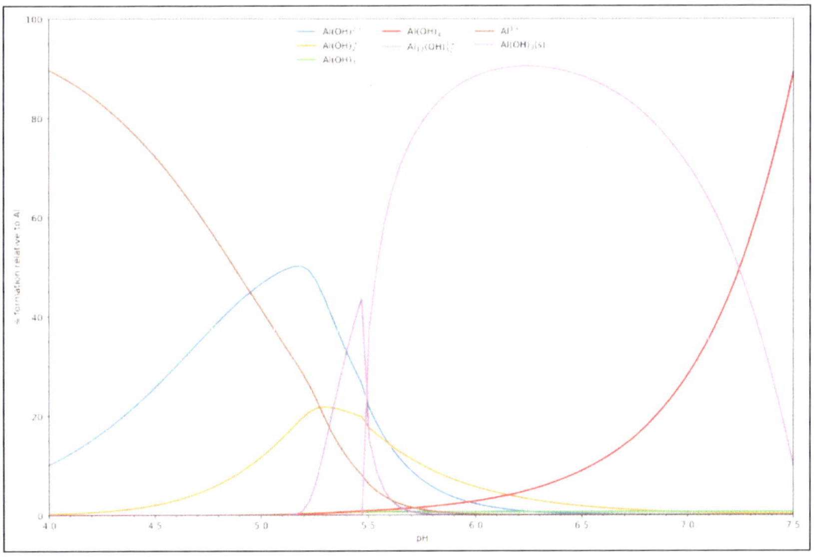

Species concentrations during hydrolysis of the aluminium.

The composition of solutions containing reactants A and H is easy to calculate as a function of p[H]. When [H] is known, the free concentration [A] is calculated from the mass-balance equation in A.

The diagram alongside, shows an example of the hydrolysis of the aluminium Lewis acid $Al^{3+}_{(aq)}$ shows the species concentrations for a 5×10^{-6} M solution of an aluminium salt as a function of pH. Each concentration is shown as a percentage of the total aluminium.

Solution and Precipitation

The diagram above illustrates the point that a precipitate that is not one of the main species in the solution equilibrium may be formed. At pH just below 5.5 the main species present in a 5 μM solution of Al^{3+} are aluminium hydroxides $Al(OH)^{2+}$, $AlOH_2^+$ and $Al_{13}(OH)_{32}^{7+}$, but on raising the pH $Al(OH)_3$ precipitates from the solution. This occurs because $Al(OH)_3$ has a very large lattice energy. As the pH rises more and more $Al(OH)_3$ comes out of solution. This is an example of Le Châtelier's principle in action: Increasing the concentration of the hydroxide ion causes more aluminium hydroxide to precipitate, which removes hydroxide from the solution. When the hydroxide concentration becomes sufficiently high the soluble aluminate, $Al(OH)^-_4$, is formed.

Another common instance where precipitation occurs is when a metal cation interacts with an anionic ligand to form an electrically neutral complex. If the complex is hydrophobic, it will precipitate out of water. This occurs with the nickel ion Ni^{2+} and dimethylglyoxime, ($dmgH_2$): in this case the lattice energy of the solid is not particularly large, but it greatly exceeds the energy of solvation of the molecule $Ni(dmgH)_2$.

Minimization of Gibbs Energy

At equilibrium, at a specified temperature and pressure, and with no external forces, the Gibbs free energy G is at a minimum:

$$dG = \sum_{j=1}^{m} \mu_j \, dN_j = 0$$

where μ_j is the chemical potential of molecular species j, and N_j is the amount of molecular species j. It may be expressed in terms of thermodynamic activity as:

$$\mu_j = \mu_j^{\ominus} + RT \ln A_j$$

where $\mu_j^{\ominus}$ is the chemical potential in the standard state, R is the gas constant T is the absolute temperature, and A_j is the activity.

For a closed system, no particles may enter or leave, although they may combine in various ways. The total number of atoms of each element will remain constant. This means that the minimization above must be subjected to the constraints:

$$\sum_{j=1}^{m} a_{ij} N_j = b_i^0$$

where a_{ij} is the number of atoms of element i in molecule j and b_i^0 is the total number of atoms of element i, which is a constant, since the system is closed. If there are a total of

k types of atoms in the system, then there will be k such equations. If ions are involved, an additional row is added to the a_{ij} matrix specifying the respective charge on each molecule which will sum to zero.

This is a standard problem in optimisation, known as constrained minimisation. The most common method of solving it is using the method of Lagrange multipliers (although other methods may be used).

$$\mathcal{G} = G + \sum_{i=1}^{k} \lambda_i \left(\sum_{j=1}^{m} a_{ij} N_j - b_i^0 \right) = 0$$

where the λ_i are the Lagrange multipliers, one for each element. This allows each of the N_j and λ_j to be treated independently, and it can be shown using the tools of multivariate calculus that the equilibrium condition is given by:

$$0 = \frac{\partial \mathcal{G}}{\partial N_j} = \mu_j + \sum_{i=1}^{k} \lambda_i a_{ij}$$

$$0 = \frac{\partial \mathcal{G}}{\partial \lambda_i} = \sum_{j=1}^{m} a_{ij} N_j - b_i^0$$

This is a set of (m + k) equations in (m + k) unknowns (the N_j and the λ_i) and may, therefore, be solved for the equilibrium concentrations N_j as long as the chemical activities are known as functions of the concentrations at the given temperature and pressure. (In the ideal case, activities are proportional to concentrations.) Note that the second equation is just the initial constraints for minimization.

This method of calculating equilibrium chemical concentrations is useful for systems with a large number of different molecules. The use of k atomic element conservation equations for the mass constraint is straightforward, and replaces the use of the stoichiometric coefficient equations. The results are consistent with those specified by chemical equations. For example, if equilibrium is specified by a single chemical equation:

$$\sum_{j=0}^{m} \nu_j R_j = 0$$

where ν_j is the stochiometric coefficient for the j th molecule (negative for reactants, positive for products) and R_j is the symbol for the j th molecule, a properly balanced equation will obey:

$$\sum_{j=1}^{m} a_{ij} \nu_j = 0$$

Multiplying the first equilibrium condition by v_j yields:

$$0 = \sum_{j=1}^{m} v_j \mu_j + \sum_{j=1}^{m}\sum_{i=1}^{k} v_j \lambda_i a_{ij} = \sum_{j=1}^{m} v_j \mu_j$$

As above, defining ΔG:

$$\Delta G = \sum_{j=1}^{m} v_j \mu_j = \sum_{j=1}^{m} v_j (\mu_j^{\ominus} + RT\ln(\{R_j\})) = \Delta G^{\ominus} + RT\ln\left(\prod_{j=1}^{m} \{R_j\}^{v_j}\right) = \Delta G^{\ominus} + RT\ln(K_c)$$

where K_c is the equilibrium constant, and ΔG will be zero at equilibrium.

Analogous procedures exist for the minimization of other thermodynamic potentials.

Collision Theory

According to the collision theory, "the molecules of reactants are assumed to be hard spheres and the reactions are assumed to occur only when these spheres (molecules) collide with each other". So it was important to quantify the number of collisions occurring in order to form products so that we can have a clear picture of the reaction, and hence came the term collision frequency.

Collision Frequency can be understood as the number of collisions per second per unit volume of the reacting mixture. It is generally denoted by Z.

Let's consider the following bimolecular elementary reaction:

$$P + Q \rightarrow \text{Product}$$

Now as per the collision theory, the rate of the above reaction can be given by:

$$Rate = Z_{PQ} e^{-\frac{Ea}{RT}}$$

Where:

- Z_{PQ} = collision frequency of reactants P and Q.
- E_a = Activation Energy.
- R = Universal Gas Constant.
- T = Temperature in absolute scale.

One more parameter that affected the rates of chemical reactions significantly is activation energy (E_a). The term activation energy was given by Arrhenius, it is the minimum amount of energy that the reactants must possess in order to form a product during a chemical reaction.

Activation Energy

The activation energy is the minimum amount of energy needed by the reacting particles in any particular reaction for that reaction to take place. Unless particles collide with sufficient energy to supply the activation energy they simply don't react. The activation energy must be supplied before a reaction can take place.

To start a chemical reaction chemical bonds need to be broken in the reactants, which requires energy. The energy needed to start the reaction is known as activation energy. Sometimes the activation energy is low so that the reaction can start at room temperature without heating.

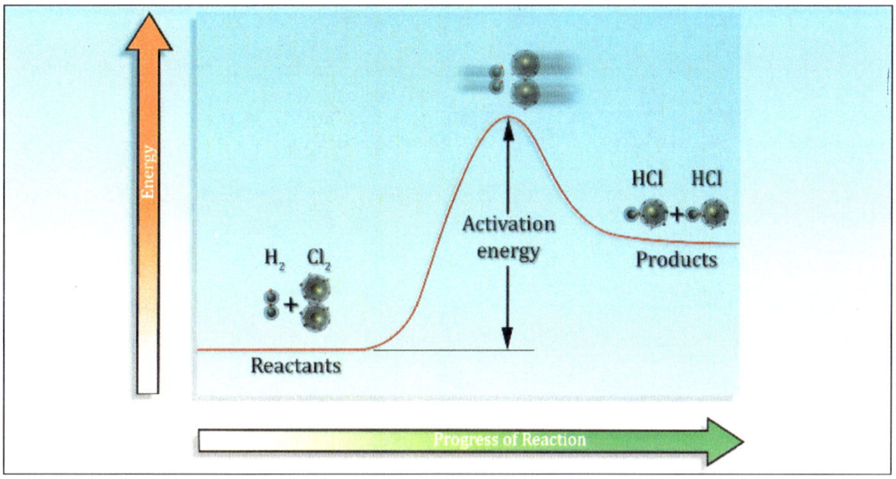

Activation Energy

Arrhenius Equation

As per the Arrhenius Equation, all the molecules that have energy greater than or equal to activation energy will collide to form products. But this postulate was not true for all the reactions. A significant amount of deviation was observed in reactions involving complex molecules.

Some of the molecules having sufficient energy (activation energy) did not collide to form the product. Only a few of them made effective collisions, collisions that led to the formation of products. The scientists found that the kinetic energy of the molecules is not the only parameter that governs the reaction.

They concluded that only those molecules that have the threshold energy (activation energy) and proper orientation during the collision will form products. They intro duced a probability factor P to account for the effective collisions.

So now the rate of a reaction is given by:

$$Rate = PZ_{PQ}e^{-\frac{Ea}{RT}}$$

Collision Theory

In a nutshell, we can say that the activation energy and proper orientation of the reacting molecules together determine the condition for a collision that will result in the formation of products that is an effective collision. So in collision theory, both activation energy and effective collision govern the rate of a reaction.

References

- Housecroft, C.E.; Sharpe, A. G. (2008). Inorganic Chemistry (3rd ed.). Prentice Hall. ISBN 0-13-175553-6. Chapter 6: Acids, Bases and Ions in Aqueous Solution
- Chemical-kinetics, science: britannica.com, Retrieved 18 April, 2019
- Skoog, D. A.; West, D. M.; Holler, J. F.; Crouch, S. R. (2004). Fundamentals of Analytical Chemistry (8th ed.). Thomson Brooks/Cole. ISBN 0-03-035523-0. Section 30E, Chromatographic separations
- Chen, Luonan; Wang, Ruiqi; Li, Chunguang; Aihara, Kazuyuki (2010). Modeling Biomolecular Networks in Cells. Doi:10.1007/978-1-84996-214-8. ISBN 978-1-84996-213-1
- Law-of-mass-action-or-law-of-chemical-equilibrium, chemistry: byjus.com, Retrieved 11 January, 2019
- Atkins, Peter; De Paula, Julio (2006). Atkins' Physical Chemistry (8th ed.). W. H. Freeman. Pp. 200–202. ISBN 0-7167-8759-8
- The-collision-theory-of-chemical-reactions, chemistry: byjus.com, Retrieved 14 July, 2019

Electrochemical Engineering 4

- **Electrochemical Reaction**
- **Standard Electrode Potential**
- **Electrochemical Cells**

The branch of chemical engineering that deals with the applications of various electrochemical phenomena such as electrowinning and refining, electrochemical separations and corrosion is known as electrochemical engineering. This chapter has been carefully written to provide an easy understanding about electrochemical engineering and electrochemical cells.

Electrochemical engineering is the knowledge required to either design and run an industrial plant which includes an electrolytic stage for the production of chemicals or to produce an electrolytic device for the generation of power. The former involves the use of electric power for the production of chemicals and the latter the use of chemicals for the production of electric power. There are also a number of processes based on electrolysis, for example, electrochemical machining and electrophoretic painting, which clearly belong to the domain of electrochemical engineering.

Electrochemical Reaction

Electrochemical reaction refers to any process either caused or accompanied by the passage of an electric current and involving in most cases the transfer of electrons between two substances—one a solid and the other a liquid.

Under ordinary conditions, the occurrence of a chemical reaction is accompanied by the liberation or absorption of heat and not of any other form of energy; but there are many chemical reactions that—when allowed to proceed in contact with two electronic conductors, separated by conducting wires—liberate what is called electrical energy, and an electric current is generated. Conversely, the energy of an electric current can be used to bring about many chemical reactions that do not occur spontaneously. A process involving the direct conversion of chemical energy when suitably organized constitutes an electrical cell. A process whereby electrical energy is converted directly into chemical energy is one of electrolysis; i.e., an electrolytic process. By virtue of their combined chemical energy, the products of an electrolytic process have a tendency to react spontaneously with one another, reproducing the substances that were reactants and were therefore consumed during the electrolysis. If this reverse reaction is allowed to occur under proper conditions, a large proportion of the electrical energy used in the electrolysis may be regenerated. This possibility is made use of in accumulators or storage cells, sets of which are known as storage batteries. The charging of an accumulator is a process of electrolysis; a chemical change is produced by the electric current passing through it. In the discharge of the cell, the reverse chemical change occurs, the accumulator acting as a cell that produces an electric current.

Principles

Substances that are reasonably good conductors of electricity may be divided into two groups: the metallic, or electronic, conductors and the electrolytic conductors. The metals and many nonmetallic substances such as graphite, manganese dioxide, and lead sulfide exhibit metallic conductivity; the passage of an electric current through them produces heating and magnetic effects but no chemical changes. Electrolytic conductors, or electrolytes, comprise most acids, bases, and salts, either in the molten condition or in solution in water or other solvents. Plates or rods composed of a suitable metallic conductor dipping into the fluid electrolyte are employed to conduct the current into and out of the liquid; i.e., to act as electrodes. When a current is passed between electrodes through an electrolyte, not only are heating and magnetic effects produced but also definite chemical changes occur. At or in the neighbourhood of the negative electrode, called the cathode, the chemical change may be the deposition of a metal or the liberation of hydrogen and formation of a basic substance or some other chemical reduction process; at the positive electrode, or anode, it may be the dissolution of the anode itself, the liberation of a nonmetal, the production of oxygen and an acidic substance, or some other chemical oxidation process.

An electrolyte, prepared either by the melting of a suitable substance or by the dissolving of it in water or other liquid, owes its characteristic properties to the presence in it of electrically charged atoms or groups of atoms produced by the spontaneous splitting up or dissociation of the molecules of the substance. In solutions of the so-called strong electrolytes, most of the original substance, or in some solutions perhaps all of it, has undergone this process of electrolytic dissociation into charged particles, or ions. When an electrical potential difference (i.e., a difference in degree of electrification) is established between electrodes dipping into an electrolyte, positively charged ions move toward the cathode and ions bearing negative charges move toward the anode. The electric current is carried through the electrolyte by this migration of the ions. When an ion reaches the electrode of opposite polarity, its electrical charge is donated to the metal, or an electric charge is received from the metal. The ion is thereby converted into an ordinary neutral atom or group of atoms. It is this discharge of ions that gives rise to one of the types of chemical changes occurring at electrodes.

Electrochemical Process

Interactions of matter associated with the passage of an electric current depend upon the characteristics of the negatively charged electron. As the basic particle of electricity, the electron has an affinity for positively charged particles of matter, protons, whether in atoms, groups of atoms, or molecules. This affinity is analogous to the chemical affinity that particles exhibit among themselves. In fact, all chemical reactions result from a shift in the electron structure of atoms, and free electrons can combine with particles of matter (reduction) or be released by them (oxidation). The quantitative relationship between the free electrons of an electric current and the particles of a substance in which they cause a reaction is defined by the laws of Faraday.

Electrons are available in large quantities in a relatively free (mobile) state only in substances called electronic conductors, among which metals are the most important. Thus, an electron conductor must be present as a basic component of any system in which electrochemical reactions are to occur. Furthermore, the availability of electrons in a conductor is limited by energy distribution to such an extent that electrochemical reactions take place only in the immediate vicinity of the electronic conductor's surface—i.e., a few angstroms from the conductor into the solution. These reactions are, therefore, normally considered as occurring at the interface, or common boundary, between an electronic conductor, such as an electrode, and an ionic conductor of electricity, such as an electrolytic solution. Electrochemical reaction will take place, however, only to the extent that electricity can flow through such a system as a whole. To achieve this, it is necessary for the system to form a closed loop, electronically speaking.

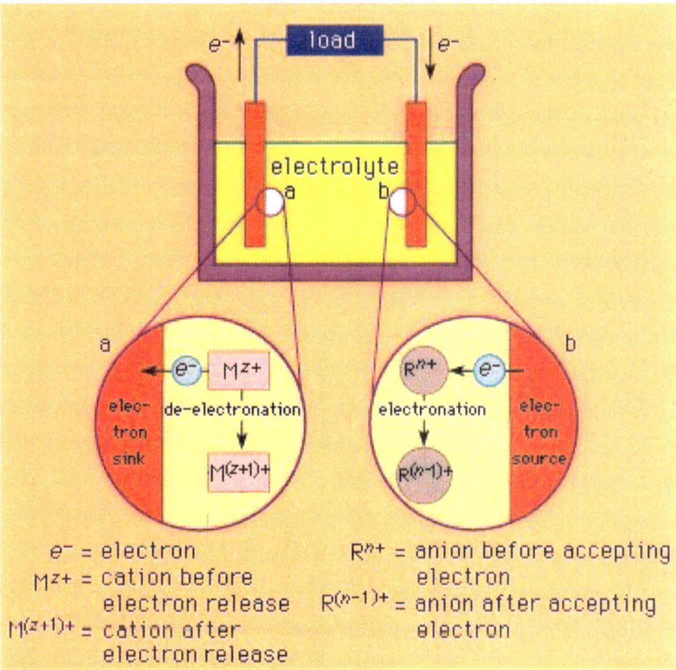

The mechanism of electron movement in an electrochemical cell.

To summarize, if at one metal-solution interface electrons are coming out of the metal, reducing a component of the solution, there must exist a second metal-solution interface where electrons are going into the metal in the process of oxidation.

The two electrodes and the ionic conductor in between (e.g., an aqueous solution of some ionized salt) represent an electrochemical cell. The process occurring in the cell as a whole is a redox process with the reduction of one species spatially separated from the oxidation of another one. As a consequence of Faraday's law, the rates of electrochemical reactions at electrodes (expressed in terms of gram moles per second per square centimetre of electrode surface) are directly proportional to the current density (expressed in amperes per square centimetre)—i.e., current flowing through the cell divided by the electrode surface area.

Sites of Electrochemical Reactions

Electrochemical reactions take place where the electron conductor meets the ionic conductor—i.e., at the electrode–electrolyte interface. Characteristic of this region, considered to be a surface phase, is the existence of a specific structure of particles and the presence of an electric field of considerable intensity (up to 10,000,000 volts per centimetre) across it; the field is caused by the separation of charges that are present between the two bulk phases in contact. For most purposes the surface phase can be considered as a parallel plate condenser, with one plate on the centre of the ions that have been brought to the electrode, at the distance of their closest approach to it, and with the second plate at the metal surface; between the two plates and acting as a dielectric (i.e., a nonconducting material) are oriented water molecules. This structure is termed the electric double layer and is illustrated in figure.

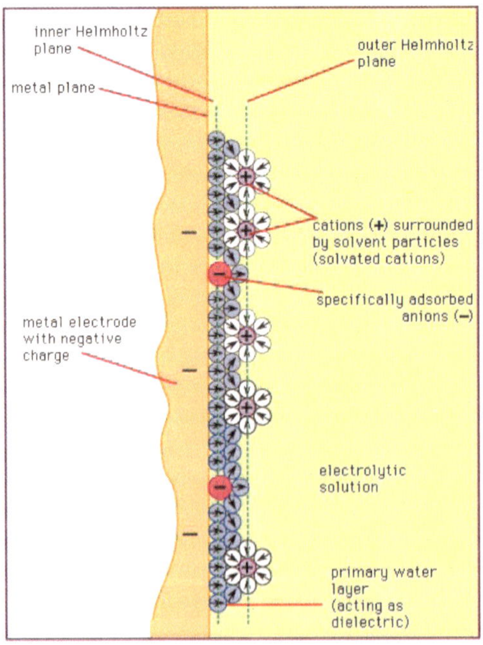

Double-layer structure and change of potential with distance from the electrode surface.

Thermal motion of the positive ions in the solution makes the condenser plate on the electrolyte side of the interface diffuse—i.e., the ions are distributed in a cloudlike way. This condition justifies the division of the potential change between the bulk of metal and the bulk of electrolyte into two parts: first, that between the metal surface and the first ionic layer at the distance of closest approach (called the outer Helmholtz plane, in which the ions are usually surrounded by solvent particles—i.e., are solvated); and second, that between the first ionic layer and the bulk of the solution, the diffuse part of the double layer. The picture is further complicated by the presence of ions in the electrode surface layer in addition to those that are present for electrostatic reasons—i.e., by the force of attraction or repulsion between electric charges. Such electrode surface layer ions are said to be specifically adsorbed on the electrode surface. Since this species of ions is attracted by the surface to a distance closer than the "distance of the closest approach" of ions, further subdivision of the inner part of the electric double layer is justified. Hence, the inner Helmholtz plane is introduced as the plane formed by the centres of specifically adsorbed ions. Adsorption of neutral molecules on the surface can also change the properties of the electric double layer. This change occurs as a consequence of replacing the water molecules, and thus changes that part of the potential (electrical) difference across the double layer that is caused by the adsorbed dipoles (water molecules that have a polarity—i.e., they behave like minute magnets—because of their hydrogen-oxygen structure, making one end of the molecule positive and the other end negative).

The absolute value of electrical potential difference, symbolized in calculation by delta and psi, $\Delta\psi$, between the bulk of a metal electrode and the bulk of an electrolyte cannot be measured. Instead, the voltage of a special cell, composed of the specific electrode being studied and of an arbitrarily selected reference electrode, is normally measured; the voltage is referred to as the relative electrode potential, E. Of special interest is that state of the electrode at which there is no net charge (in this case, no unbalanced, or extra positive, charge) at the metal side of the double layer. The relative potential at which this state is achieved is characteristic of each metal. This point is termed the potential of zero charge. At that potential, the field across the double layer is due to orientation of water molecules and other dipoles at the surface only.

Most of the knowledge of the detailed structure of the interface between a metal and an electrolyte arises from experimentation with mercury, the only metal that is liquid at ordinary temperatures; the double layer structure turns out to have surface tensions that must be measured, and this measurement is difficult with solid metals. By 1970, however, it had been shown that it is possible to measure surface tension changes at the metal-solution interface. Thus, the way to the determination of the double layer structure involving solids was opened.

Substances that are semiconductors can also be employed as electron carriers in electrochemical reactions. Semiconductors are substances which range between serving as insulators at low temperatures and as metallic-type conductors at high temperatures.

In the case of semiconductors, however, the electric double layer has a more complex structure inasmuch as the condenser plate at the electrode side of the double layer also becomes diffuse. Thus, the overall potential difference between bulks of the phases in contact comprises also the potential difference between the bulk of the semiconductor and its surface.

Types of Electrochemical Reactions

There are several types of electrochemical reactions.

Simple Redox Reactions

A simple redox reaction is one that involves a change in the electrical charge of a charge carrier, usually a simple or complex ion in the solution, by its taking away, an electron from the electrode (reduction), or its giving an electron to the electrode (oxidation). The same carrier may be present in solution in two states of charge. The higher, more positive charge is called the oxidized state, and the lower, less positive charge is called the reduced state. For example, when ferric and ferrous ions are both present in solution in significant quantity, and when electron exchange with the electrode is sufficiently fast, redox equilibrium is established at the electrode, giving it a well-defined potential, or reversible redox potential.

Reactions that Produce Gases

When hydrogen ions in solution react with electrons ejected from a metal, hydrogen atoms are formed at the surface, where they combine among themselves or with other hydrogen ions and electrons to give gaseous hydrogen molecules. If all the reactions are fast enough, an equilibrium is attained between hydrogen ions and gaseous hydrogen. A metal in contact with solution at which such a situation exists is called the reversible hydrogen electrode, and its electrical potential is arbitrarily taken to be zero; every other electrode can thus be compared with it as it represents the basis for constituting the hydrogen scale of relative electrode potentials. Similarly, negative hydroxyl ions in solution (OH^-) can be made to give up electrons to a metal and, in a series of reactions, the final one is the formation of gaseous oxygen. Chlorine is another gaseous product; it evolves upon electrochemical oxidation of chloride ions in concentrated solutions of neutral and acid salts.

Reactions that Deposit and Dissolve Metals

When a metal ion is reduced and discharged as a neutral atom, or species, it tends to build into the metal lattice of the electrode. Thus, metals can be deposited at electrodes. Conversely, if electrons are taken away from the metal electrode by applying positive potentials to it, the metal ions thus formed can cross the double layer of electric charge at the interface, undergo hydration (combination with water), and enter the solution.

The metal electrode thus dissolves. Many metals establish well-defined electric potentials when they are in contact with their own ions in solution.

Oxidation and Reduction of Organic Compounds

A reaction of the oxidation and reduction of organic compounds can also be done at electrodes. Such reactions, however, are mostly irreversible in the literal sense that they lead to products that cannot easily be converted back into the original substance. Exceptions are some oxygen- and nitrogen-containing compounds (quinones, amines, and nitrous compounds) that can give fairly well-defined reversible potentials.

Mechanism of Charge Transfer

The causes of the thermodynamically irreversible behaviour of electrode reactions are found in the nature of the elementary act of charge transfer. Like any chemical reaction, this act is inhibited by the existence of an energy barrier between the oxidized and the reduced state. This barrier implies that the reaction could take place only in the special circumstances when, during the course of numerous interactions with other species (atoms, ions, molecules, etc.) surrounding it, a molecule attains an excited state in which it has an abnormal energy content. In most chemical reactions, this energy content must be sufficient for the species to come into what is called the transition state; the transition state characterizes the top of the energy barrier just before a reaction begins. If such a model is applied to electron transfer at an interface, calculation shows that electron exchange reactions at electrodes would be prohibitively slow, a conclusion at variance with the observed phenomena; quantum mechanical laws, however, govern the motion of electrons, and their inclusion changes the calculations to fit reality. Quantum mechanics require that for fast electron exchange to take place, electrons in a particle outside the double layer (e.g., a hydrated ion at the outer Helmholtz plane) must attain certain well-defined quantized energy levels equal to those in which free electrons exist in the metal. Since such states can be attained by the particle at a lower energy-content than that needed for its transfer over the top of the energy barrier, according to the classical view, this fast process of electron exchange between the electrode and a particle in solution is termed electron tunnelling through the energy barrier.

Whereas the rate of chemical processes, or what may loosely be termed the speed of reaction, can be influenced only by changing the concentrations of reactants or by changing the temperature or both, the rate of electrochemical processes also can be manipulated by changing the electrode potential. Making the electrode more negative increases the number of electrons in the metal ready to tunnel to ions, and hence the rate of the reduction process increases. Conversely, making the potential more positive decreases this rate and increases the number of particles ready to give away electrons, thus increasing the rate of the oxidation process.

It can be deduced that there must exist a direct proportionality between the rate of reaction and the concentration of the reacting species and at the same time an exponential proportionality between the rate of reaction and the electrode potential.

At any electrode potential, both reduction of one species and oxidation of the product of reduction are taking place but at different rates; the rate of each reaction is determined by the respective concentration and by the corresponding effects of potential. The rate of an electrochemical reaction can best be described as the electric current density—i.e., a measure of the quantity of electrons moving in a certain volume of space during a specified unit of time.

The relationships can be represented quantitatively by an equation in which the net, or resulting, current (the difference of the rate of electron ejection across the interface to particles in solution, diminished by the rate at which particles in solution inject electrons into the metal) is equated to the difference of the rates of reduction and oxidation and the variables and constants that relate to these reactions. Equation:

$$E_{\text{rev}} = \frac{RT}{(\alpha_c + \alpha_a)F} \ln \frac{\vec{k}}{\overleftarrow{k}} + \frac{RT}{(\alpha_c + \alpha_a)F} \ln \frac{C_{\text{ox}}}{C_{\text{red}}}$$
$$= E^0 + \frac{RT}{nF} \ln \frac{C_{\text{ox}}}{C_{\text{red}}}.$$

known as the Nernst equation, which can be derived from equation:

$$i = \vec{i} - \overleftarrow{i} = \vec{k} C_{\text{red}} \exp\left(\frac{\alpha_a F}{RT} E\right) - \overleftarrow{k} C_{\text{ox}} \exp\left(\frac{\alpha_c F}{RT} E\right),$$

gives the value of the electrode potential when the rate of oxidation exactly equals the rate of reduction. Using this value for the electrode potential, equation

$$i = i_0 \left\{ \frac{(C_{\text{red}})i}{(C_{\text{red}})_0} \exp\left(\frac{\alpha_a F \eta}{RT}\right) - \frac{(C_{\text{ox}})i}{(C_{\text{ox}})_0} \exp\left(\frac{\alpha_c F \eta}{RT}\right) \right\},$$ called the Butler-Volmer equation,

can be derived; it represents the most fundamental relationship in electrodic chemistry.

Complex Electrochemical Reactions

Electrochemical processes considered so far involve simple reactions of a particle with a single electron to produce a reduced ion (e.g., the ferrous ion of iron with two positive charges, Fe^{++}), or vice versa. Such are the simple ionic redox processes, where the only difference in structure between a reactant and a reaction product may be due to some rearrangement of the neighbouring solvent molecules. When one or more transfers of electrons between the electrode and a species in solution are accompanied by major structural changes (e.g., when hydroxyl ions, OH^-, transform into a molecule of oxygen, O_2, and a water molecule, H_2O, in the process of oxygen evolution at the anode, or positive electrode), the reaction usually consists of a sequence of events, called elementary acts, or unit steps, constituting the reaction mechanism. Intermediate states between the steps usually involve some unstable intermediate species with higher energy content than those of the reactants or of the reaction products.

Complex reaction mechanisms can consist of a number of electron transfer steps, with some chemical steps preceding or succeeding the electron transfer steps or taking place in between them. Most organic electrochemical reactions are complex, involving large numbers of electrons in the overall reaction. Usually one step in the reaction encounters the largest energy barrier. The rate of occurrence of this step limits the rate of the overall reaction (i.e., all other steps must occur at the same net rate, although they could provide for a much faster overall change). This step is called the rate-determining step and, for most practical purposes, all intermediate steps before and after it can be considered to be in equilibrium. It is interesting to note that whenever this is the case, the Butler-Volmer equation is applicable, but with specific values of the transfer coefficients αa and αc characteristic of the mechanism of the reaction.

Experimental Studies

Measuring the rates of electrochemical reactions (i.e., current densities) as functions of electrode potential under steady-state conditions represents the normal tool of electrodics. Meaningful results could not be obtained, however, until the sensitivity of electrochemical reactions to impurities was realized and high purity techniques were introduced. Even so, the steady-state method often has shortcomings except for relatively slow electrode reactions. In many cases concentration changes at the electrodes prevent using a sufficiently wide current density range for obtaining meaningful Tafel relationships. Hence, so-called transient methods have been developed in which one electrochemical factor in the situation is rationally perturbed and the time dependence of others observed. One such method consists of placing a constant current pulse upon an electrode and measuring the variation of the resulting current through the solution. This is called the galvanostatic method for measuring the rate of an electrochemical reaction. Applying a potential pulse while observing the variation of the rate as a function of time constitutes the potentiostatic method. A third method, called the potentiodynamic, or potential sweep, method involves observations of the current as a function of the potential, while the latter is varied at a constant, known rate.

The advantages of transient methods over steady-state ones, in which behaviour before the attainment of the steady-state is not part of the observation, are manifold. If observations are made at sufficiently short times, events can be recorded before the onset of concentration changes, and pure activation values can be found. Hence, Tafel relationships can be obtained over a larger current density range than if one makes measurements over longer times, as is required in the steady-state methods. The structure of the transient states can reveal important information, such as double layer capacitance and surface coverage of the electrode by intermediate species.

Several so-called kinetic parameters, for example, partial derivatives of current density and potential with respect to concentration of chosen reactants, can be extracted from experimental measurements.

Electrochemical measurements have a limited capacity to reveal the state of an electrode surface. Nonelectrochemical methods of studying electrode surfaces, therefore, have been stressed. Optical methods have considerably gained in importance. Ellipsometry (i.e., measuring changes in basic properties of polarized light as it is reflected from an electrode surface) was the first method that made possible a study of monomolecular layers of oxides and adsorbed oxygen as adsorbed organic molecules. Adaptation of such a method to transient use allows change in the surface to be related to the passivity of metals. Another kind of spectroscopy enables infrared spectra of species adsorbed at electrode surface to be taken. Mössbauer spectra may lead to an identification of thin layers on an electrode surface. A good future is seen for further development of refined techniques for the study of electrochemical processes by various combinations of spectroscopic and electrochemical means.

Types of Complex Reactions

Electrocatalysis

The problems related to the increase of rates of electrochemical reactions, or, to put it another way, the decrease of overpotential, needed to perform reactions at a given rate are the subject of electrocatalysis. Both increase and decrease are of considerable practical importance since they affect the economics of electrochemical processes. Electrocatalysis is concerned with the electrode as a substrate, or base, for electrochemical reaction and with the effect of its bulk and surface properties on the rate of reaction. Contrary to expectation, the rate of the basic step in electrochemical reaction—electron transfer—is independent of the ease with which electrons are released from the metal. Hence, those simple electrode processes in which no other changes but electron transfer take place have heats of activation virtually independent of the metal substrate.

In a reaction involving the formation of a chemical bond between the electrode substrate and one of the radicals (charged particles) formed on the surface, rates of reaction at a given potential may vary for different substrates by many orders of magnitude. The variation is a function of the strength (or energy) of bonds established between the intermediate species and the surface, and when the substrates are transition metals, regular relations between reaction rate and certain characteristics of the metal's electronic structure are observed. If such bonds are too weak, the catalytic effect of the substrate is small. If they are too strong, the intermediate species is too stable to react further, and the surface of the metal becomes virtually blocked for further reaction. Thus, there is an optimum bond strength that can be seen if the rate of a reaction is plotted against a property of the metal substrate, which is proportional to the ability of the surface to form bonds—e.g., its heat of sublimation.

Electrochemical reactions can also be catalyzed or inhibited by foreign species present in the electrolytic solution. In the most general case, ions can change the rates of reactions by changing the properties of the double layer by specific adsorption effects. Organic molecules inhibit electrochemical reaction by blocking the surface. For some organic ions or dipoles, the blocking effect can begin at a certain well-defined potential. Before attaining it, the electrode reaction gives the usual current density-potential relationship, which is then suddenly interrupted by an abnormal fall of the current, corresponding to adsorption.

Electrocrystallization

Deposition of metals and other substances at electrodes as a consequence of an electrode process exhibits a number of specific features. The electrode process is followed by crystal building, and this results in a continuous change of the electrode surface. This change, in turn, affects the electrochemical properties of the system—the double-layer capacity and the rate constants of the charge transfer processes. Hence, if electrochemical properties are to be studied, transient methods should be employed that allow measurements to be made before major changes in surface morphology (structure) take place.

Since the two steps, the discharge of ions at the electrode and the incorporation of the discharged ions into the crystal lattice, are separated in time and space, an intermediate species exists at the surface, that of relatively loosely bound and freely moving atoms, called adatoms. Since the electrons tend to join the rest in the bulk of the metal, adatoms appear to have a partial charge, less than that of the elementary positive charge. The adatoms therefore attract solvent molecules, and the species is partially solvated. This reaction justifies considering an adatom as a kind of adsorbed ion, called an adion, which, however, has already undergone partial discharge.

Because such an intermediate state is possible, ions need not be reduced to the neutral atomic state at the point of incorporation into the crystal lattice. Indeed, energy considerations reveal this reduction to be improbable. Instead, discharge of ions is favoured on crystal planes. Their motion from the location of discharge to a site on the crystal occurs by surface diffusion. When such surface diffusion is inhibited, it can, in some cases, control the rate of reaction.

Problems of the crystal-building step in electrocrystallization are in many respects identical with those of crystal growth from the gas phase. Crystal building can occur in any one of three possible ways.

Kinks, naturally existing at any metal surface, form a suitable half-lattice position at which an atom is surrounded by one-half of the number of atoms that would surround it in the bulk of the metal; there, adatoms can be successively trapped and thus the crystal lattice is extended along a crystal edge and further on across the surface. This step growth mechanism is shown in figure. The mechanism, however, has a limited capacity for crystal growth. A step can move as far as the edge of a crystal, and step growth would lead to smoothing of the surface to perfection, but then further growth would cease.

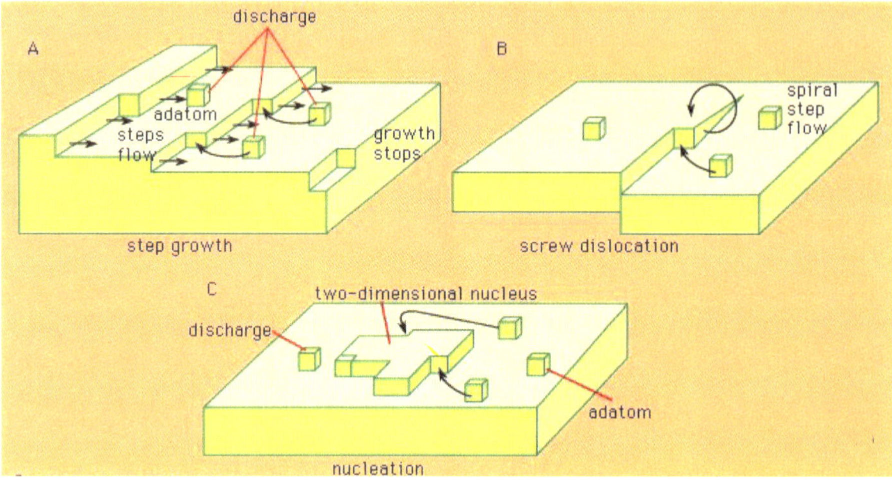

Mechanisms of crystal growth.

Mechanisms associated with screw dislocations, or twinning edges, can provide for a continuous growth of crystals. The screw dislocation mechanism, shown in figure, is made possible by a specific fault often found in the crystal lattice that may be called a dislocation originating from a shift of one atom in the lattice with respect to a perfect arrangement. This shift may then result in the formation of a monoatomic edge projecting above the electrode surface at which new atoms can be stacked. The stacking produces a turning of the edge around the base atom as a centre, the process producing a spiral growth.

If growth sites are rare, or if the substrate at which deposition should take place is foreign to the depositing metal, the charge transfer results in an accumulation of adatoms to a concentration considerably larger than that which can exist there at equilibrium with the crystal lattice. In such a situation, termed supersaturation, agglomeration of adatoms to form crystal nuclei is favoured. Surface energy requirements show that, at any degree of supersaturation, nuclei of certain dimensions are stable and can represent sites for further growth, as shown in figure.

The formation of deposits is controlled by the above mechanism as long as the discharge process supplies ample amounts of adatoms as building material over the entire surface. If the rate of deposition is increased, so as to produce near the surface considerable depletion of ions, uneven deposition will start. This is caused by the protrusions of a normally rough metal surface being closer to the bulk of solution than the recessed parts and, hence, getting a somewhat faster supply of the discharging species. Once such a situation is established, it tends to develop further. The faster growing points penetrate into ever richer layers of solution, resulting in ever faster growth. Thus, a natural consequence of deposition under transport-controlled rate is the amplification of the original surface roughness and the appearance of protruding spikes, called dendrites, as deposits.

A converse process is the smoothing of the original surface irregularities, which may occur when some foreign species, called an additive, is present in solution and adsorbs on the surface and inhibits the process of discharge. If those molecules are incorporated into a growing deposit, a situation may arise in which their supply to recessed parts of the surface becomes slower than to elevated parts. As a result, deposition becomes faster at recessed parts than at elevated ones and leveling of the surface occurs. This process has considerable technical application. Metal dissolution can sometimes be governed by a similar, transport-controlled mechanism, in which case a polishing effect is obtained. Electropolishing has found wide application in practice.

Organic Electrode Reactions

A very large number of electrochemical reactions involving organic molecules are known. An example is the oxidation of ethylene according to the equation:

$$\underset{\text{(ethylene)}}{C_2H_4} + \underset{\text{(water)}}{4H_2O} \rightarrow \underset{\text{(carbon dioxide)}}{2CO_2} + \underset{\text{(hydrogen ion)}}{12H^+} + \underset{\text{(electrons)}}{12e^-}$$

Many chemical organic reactions can be made to function electrochemically, a general advantage being that the rate of the reaction is easily controlled by controlling the potential. A change of potential may change the path of the reaction and hence a certain product may be tuned in. In particular, polymerization reactions at electrodes can be stimulated; such a method is used as a step in the production of nylon.

Multielectrode Systems

So far, systems have been considered in which a single electrode process takes place. In principle, at any electrode potential all species present in the system fall into two categories: those that are stable, and those that undergo oxidation or reduction. The stable species are those that at the given electrode potential would not decrease their free energy by giving off or accepting electrons.

At any potential there should occur a codischarge of all unstable species. Thus, the system can be considered as a multielectrode system, consisting of as many electrodes as there are redox couples present. The rate at which different processes occur, however, can be so widely different that usually a single process is by far the dominant one. Systems at which two electrode processes occur at comparable rates are of considerable importance. If two kinds of metal ions are discharged simultaneously, an alloy is formed upon crystallization. The properties of the alloy would in most cases be those determined by the phase diagram (a plot of the temperature of melting versus the composition of the mixed system) for the ratio of quantities of discharged metals given by the rates at which the discharges take place. In many cases, nonequilibrium metal phases are formed giving unusual properties to the alloy.

If a process of metal dissolution (an oxidation process) can occur at a rate comparable to that of some reduction process on the same metal, a corrosion couple is established. Thus, zinc immersed in acid solution tends to establish a potential sufficiently positive for the metal to eject metal ions into the solution (oxidation). At the same time, it is also sufficiently negative for the reduction of hydrogen ions, present in any aqueous solution, to hydrogen gas. Hence, a spontaneous process of hydrogen evolution and of dissolution of the metal will take place. The rate at which this corrosion process occurs is governed by rate laws of the type given by equation $i = i_0 \left\{ \frac{(C_{\text{red}})i}{C_{\text{red}})0} \exp\left(\frac{\alpha_a F\eta}{RT} \right) - \frac{(C_{\text{ox}})i}{(C_{\text{ox}})_0} \exp\left(\frac{\alpha_c F\eta}{RT} \right) \right\}$. The mixed potential, spontaneously established by the corroding metal, is obtained by equating expressions for the anodic and cathodic currents of the two processes in the corrosion couple.

Corrosion can be prevented in two ways. One is by using an external source to make the potential of the metal sufficiently negative to bring it into the potential region in which the metal is stable, called cathodic protection; and the other is by provoking by some means the formation of a film on the surface that would slow the process. Such films could consist of an oxide or a layer of organic molecules that prevents dissolution and hence is called an inhibitor.

Applications

Electrochemical processes are used in many ways and their use is likely to increase because they can replace polluting chemical situations with nonpolluting electrochemical ones. In many fields, however, applications have been profitable for some time. Major categories are listed below.

Metallurgy

All technologically important metals, except iron and steel, are either obtained or refined by electrochemical processes; for example, aluminum, titanium, alkaline earth, and alkali metals are obtained by electrodeposition from molten salts, and copper is refined by electrolysis in aqueous copper sulfate solutions.

Electroplating

One of the major ways of both decorating objects and improving their resistance to corrosion is by electroplating them. All major metal-working industries, particularly the automobile industry, have large electroplating plants.

Chemical Industry

Electrolysis of brine to obtain chlorine and caustic soda is an electrochemical process that has become one of the largest volume productions in the chemical industry. Modern processes cover a wide field, from the production of a variety of inorganic compounds to the production of such synthetic fibres as nylon. Intensive research in organic electrochemistry promises major developments in application, particularly with the prospect of greatly reduced electricity costs expected eventually to arise from the development of controlled fusion.

Batteries

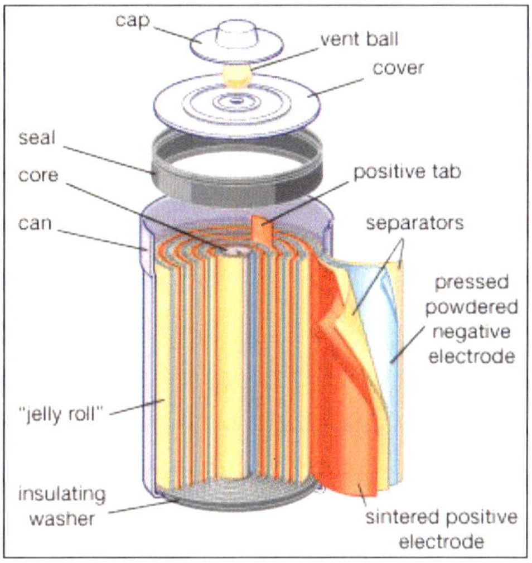

Nickel (hydroxide)-cadmium cell of "jelly roll" construction.
This rechargable battery is commonly used in portable devices.

Electrochemical storage of electricity is effected in batteries. Such devices are electrochemical cells and consist of two electrodes per unit. As the electricity to be stored is accepted on the plates of the cell, it converts substances on the plates to new substances having a higher energy than the old ones. When it is desired to make the electricity available again, the terminals of the battery are connected to the load and the substances on the battery plates retransform themselves to those originally present, giving off electricity as a product of their electrochemical reactions. The steadily rising production of the lead-acid battery is largely the result of its use for starting the internal-combustion engine, which has had an equally steady rise. Other electrochemical systems are also used as storers. The nickel-iron (Edison cell) and nickel-cadmium battery with alkaline electrolyte are both used in applications where longer lives than those of the lead-acid battery are needed; the silver-zinc battery is used to start airplane engines because of its high power per unit of weight. A variety of new systems is being investigated for covering other needs. One of the greatest challenges to electrochemists and electrochemical engineers is that of producing a battery with sufficient power and energy density to run an automobile the way gasoline (petrol) does. Even if the best hypothetical predictions for removal of polluting chemicals from automobile exhausts is realized, the cleanup will not be sufficient because the expected growth of the automobile population will continue to increase the pollutant rate.

Fuel Cells

The energy of chemical reactions is converted into electrical energy in fuel cells. In these, the fuel (e.g., hydrogen, hydrazine) is fed continuously to one electrode, while oxygen from the air is reacting at the other one. The efficiency of energy conversion in fuel cells is more than twice that attainable by conventional means—for example, by means of internal combustion.

Analytical Chemistry

In analytical chemistry, most modern automated instrumental analysis is based on electrode processes—for example, potentiometry, used to measure ionization constant.

Biological Research

In biology the idea that many biological processes, from blood clotting to the transfer of nerve impulses, are electrochemical in nature continues to spread. The biological conversion of the chemical energy of food to mechanical energy takes place at an efficiency so high that it is difficult to explain without electrochemical mechanisms. Intensive research is developing in various directions in bioelectrochemistry.

Calculations

The three equations referred to above are stated in this section, and other mathematical considerations are also included. The rate of an electrochemical reaction in terms of oxidation and reduction reactions, the concentration of the reacting species, the electrode potentials and the current densities can all be related quantitatively according to equation:

$$i = \vec{i} - \overleftarrow{i} = \vec{k}C_{\text{red}} \exp\left(\frac{\alpha_a F}{RT} E\right) - \overleftarrow{k}C_{\text{ox}} \exp\left(\frac{\alpha_c F}{RT} E\right)$$

in which i is the net current density, i→ and i← are the partial current densities of the oxidation and reduction respectively, C_{red} and C_{ox} are the concentrations of the reducing and oxidizing agents, respectively, k→ and k← are the corresponding rate constants, while αa and αc are the so-called transfer coefficients—that is, specific constants giving a proper influence factor to the exponential dependence of the rate on the potential, E. In the case of a simple one-electron transfer, these factors are termed symmetry factors, for they, in a way, reflect the symmetry of the energy barrier. It can be proved that αc + αa = n, the number of electrons exchanged in a single act of an electrode reaction.

The Nernst Equation

For a particular value of E the two partial current densities must become equal. This value of potential is the reversible electrode potential. From equation above one can deduce equation:

$$E_{rev} = \frac{RT}{(\alpha_c + \alpha_a)F} \ln \frac{\vec{k}}{\overleftarrow{k}} + \frac{RT}{(\alpha_c + \alpha_a)F} \ln \frac{C_{ox}}{C_{red}}$$
$$= E^0 + \frac{RT}{nF} \ln \frac{C_{ox}}{C_{red}}.$$

This equation is known as the Nernst equation; E° is the standard electrode potential (at $C_{ox} = C_{red} = 1$) characteristic of the given redox couple.

The standard electrode potential on the hydrogen scale is related to the thermodynamics of the electrode process. It reflects the standard free energy change of the redox reaction between the electron and the given redox couple, relative to the free energy change that takes place in the hydrogen electrode process.

The Butler-volmer Equation

The reversible electrode potential can be introduced into equation $i = \vec{i} - \overleftarrow{i} = \vec{k}C_{red} \exp\left(\frac{\alpha_a F}{RT} E\right) - \overleftarrow{k}C_{ox} \exp\left(\frac{\alpha_c F}{RT} E\right)$ and the potentials taken relative to its value.

When so expressed, they are termed overpotentials and can be stated as $\eta = E - E_{rev}$; equation $i = \vec{i} - \overleftarrow{i} - \vec{k}C_{red} \exp\left(\frac{\alpha_a F}{RT} E\right) \overleftarrow{k}C_{ox} \exp\left(\frac{\alpha_c F}{RT} E\right)$ then transforms to equation $i = i_0 \left\{ \frac{(C_{red})i}{C_{red})0} \exp\left(\frac{\alpha_a F\eta}{RT}\right) - \frac{(C_{ox})i}{(C_{ox})_0} \exp\left(\frac{\alpha_c F\eta}{RT}\right) \right\}$:

$$i = i_0 \left\{ \frac{(C_{red})i}{C_{red})0} \exp\left(\frac{\alpha_a F\eta}{RT}\right) - \frac{(C_{ox})i}{(C_{ox})_0} \exp\left(\frac{\alpha_c F\eta}{RT}\right) \right\}$$

in which i_0 represents the value of either of the (equal) electron-emitting and electron-accepting partial current densities at the reversible potential and is termed the exchange current density. Equation $i = i_0 \left\{ \frac{(C_{red})i}{C_{red})0} \exp\left(\frac{\alpha_a F\eta}{RT}\right) - \frac{(C_{ox})i}{(C_{ox})_0} \exp\left(\frac{\alpha_c F\eta}{RT}\right) \right\}$ is called the Butler-Volmer equation and represents one of the most fundamental relationships of electrochemistry.

As overpotentials, either positive or negative, become larger than about 5×10^{-2} volts (V), the second or the first term of equation $i = i_0 \left\{ \frac{(C_{red})i}{C_{red})0} \exp\left(\frac{\alpha_a F\eta}{RT}\right) - \frac{(C_{ox})i}{(C_{ox})_0} \exp\left(\frac{\alpha_c F\eta}{RT}\right) \right\}$ becomes negligible, respectively. Hence, simple exponential relationships between current (i.e., rate) and overpotential are obtained, or the overpotential can be considered as logarithmically dependent on the current density. This theoretical result is in agreement with the experimental findings of the German physical chemist Julius Tafel, and the usual plots of overpotential versus log current density are known as Tafel lines. The slope of a Tafel plot reveals the value of the transfer coefficient α for the given direction of the electrode reaction.

Difficulties in Transport of Reaction Species

The above conclusions about the overpotential-current density relationship are valid as long as the ratios of concentrations at the electrode surface of the species involved at current density i, C_i and, in the absence of current, Co, stay close to unity. As the current density is increased, the concentration gradient needed to maintain a corresponding diffusion flux of the species concerned must begin to become appreciable. This condition is possible only if the concentration of the species at the surface starts to differ appreciably from the bulk value; i.e., $(C_i)i/(C_i)o \neq 1$. The change in concentration of the discharging species at the electrode surface with time can, in principle, be obtained by using a second order partial differential equation (Fick's law), which, however, has explicit solutions only for a limited number of well-defined boundary conditions.

When significant concentration changes set in, no more exponential dependence of current density on potential can be obtained. It can be derived that, instead, a transition toward a limiting value takes place.

The important case is that in which the concentration of the discharging species at the electrode surface becomes equal to zero. The steady-state (i.e., independent of time) current density obtained in such a case is the highest possible for the given set of conditions (diffusion limiting current density). The value of the concentration gradient in this case is directly proportional to the bulk concentration of the species involved and inversely proportional to the thickness of the diffusion layer (i.e., the layer close to the electrode in which the concentration of the species differs from the species concentration in the bulk). This layer most often has a thickness fixed by hydrodynamic conditions in the solution surrounding the electrode. The definition used most often for the diffusion layer thickness is that of the German physical chemist Walther Hermann Nernst, according to whom this quantity is equal to the distance from the electrode at which the concentration would reach the bulk value if the concentration gradient were constant and equal to that at the electrode surface.

If a current larger than the limiting current is forced upon the electrode, the given electrode process will be able to sustain it only in the initial period in which the layer of solution close to the electrode is not completely exhausted of the discharging ions. As the concentration of ions tends toward zero, the electrode potential will change and another electrode process will start. The time at which the abrupt change of potential toward a new process takes place is termed the transition time. The relationship between transition time, current density, and concentration of the discharging species is given by Sand's equation:

$$\tau^{1/2} = \frac{\tau^{1/2} nFC_0 D^{1/2}}{2i}.$$

Since τ is a well-defined function of the concentration of the discharging species, the observation of the transition time can also be used as an analytical tool (chronopotentiometry).

Standard Electrode Potential

In electrochemistry, standard electrode potential is defined as the measure of the individual potential of reversible electrode at standard state with ions at an effective concentration of 1mol dm-3 at the pressure of 1 atm.

The basis for an electrochemical cell, such as the galvanic cell, is always a redox reaction which can be broken down into two half-reactions: oxidation at anode (loss of electron) and reduction at cathode (gain of electron). Electricity is generated due to electric potential difference between two electrodes. This potential difference is created as a result of the difference between individual potentials of the two metal electrodes with respect to the electrolyte. (Reversible electrode is an electrode that owes its potential to changes of a reversible nature, in contrast to electrodes used in electroplating which are destroyed during their use.) It is the measure of reducing power of any element or compound.

Although the overall potential of a cell can be measured, there is no simple way to accurately measure the electrode/electrolyte potentials in isolation. The electric potential also varies with temperature, concentration and pressure. Since the oxidation potential of a half-reaction is the negative of the reduction potential in a redox reaction, it is sufficient to calculate either one of the potentials. Therefore, standard electrode potential is commonly written as standard reduction potential.

The electrode potential cannot be obtained empirically. The galvanic cell potential results from a *pair* of electrodes. Thus, only one empirical value is available in a pair of electrodes and it is not possible to determine the value for each electrode in the pair using the empirically obtained galvanic cell potential. A reference electrode, standard hydrogen electrode (SHE), for which the potential is *defined* or agreed upon by convention, needed to be established. In this case the standard hydrogen electrode is set to 0.00 V and any electrode, for which the electrode potential is not yet known, can be paired with standard hydrogen electrode—to form a galvanic cell—and the galvanic cell potential gives the unknown electrode's potential. Using this process, any electrode with an unknown potential can be paired with either the standard hydrogen electrode or another electrode for which the potential has already been derived and that unknown value can be established.

Since the electrode potentials are conventionally defined as reduction potentials, the sign of the potential for the metal electrode being oxidized must be reversed when calculating the overall cell potential. The electrode potentials are independent of the number of electrons transferred —they are expressed in volts, which measure energy per electron transferred—and so the two electrode potentials can be simply combined to give the overall *cell* potential even if different numbers of electrons are involved in the two electrode reactions.

For practical measurements, the electrode in question is connected to the positive terminal of the electrometer, while the standard hydrogen electrode is connected to the negative terminal.

Standard Reduction Potential Table

The larger the value of the standard reduction potentials, the easier it is for the element to be reduced (accept electrons); in other words, they are better oxidizing agents. For example, F_2 has 2.87 V and Li^+ has −3.05 V. F reduces easily and is therefore a good oxidizing agent. In contrast, $Li_{(s)}$ would rather undergo oxidation (hence a good reducing agent). Thus Zn^{2+} whose standard reduction potential is −0.76 V can be oxidized by any other electrode whose standard reduction potential is greater than −0.76 V (e.g. H^+(0 V), Cu^{2+}(0.34 V), F_2(2.87 V)) and can be reduced by any electrode with standard reduction potential less than −0.76 V (e.g. H_2(−2.23 V), Na^+(−2.71 V), Li^+(−3.05 V)).

In a galvanic cell, where a spontaneous redox reaction drives the cell to produce an electric potential, Gibbs free energy $\Delta G°$ must be negative, in accordance with the following equation:

$$\Delta G°_{cell} = -nFE°_{cell}$$

where n is number of moles of electrons per mole of products and F is the Faraday constant, ~96485 C/mol. As such, the following rules apply:

If $E°_{cell} > 0$, then the process is spontaneous (galvanic cell)

If $E°_{cell} < 0$, then the process is nonspontaneous (electrolytic cell)

Thus in order to have a spontaneous reaction ($\Delta G° < 0$), $E°_{cell}$ must be positive, where:

$$E°_{cell} = E°_{cathode} - E°_{anode}$$

where E^o_{anode} is the standard potential at the anode and $E^o_{cathode}$ is the standard potential at the cathode as given in the table of standard electrode potential.

Electrochemical Cells

An electrochemical cell is a device capable of either generating electrical energy from chemical reactions or using electrical energy to cause chemical reactions. An extremely important class of oxidation and reduction reactions are used to provide useful electrical energy in batteries. A simple electrochemical cell can be made from copper and zinc metals with solutions of their sulfates. In the process of the reaction, electrons can be transferred from the zinc to the copper through an electrically conducting path as a useful electric current.

An electrochemical cell can be created by placing metallic electrodes into an electrolyte where a chemical reaction either uses or generates an electric current. Electrochemical cells which generate an electric current are called voltaic cells or galvanic cells, and common batteries consist of one or more such cells. In other electrochemical cells an externally supplied electric current is used to drive a chemical reaction which would not occur spontaneously. Such cells are called electrolytic cells.

Electrolytic Cell

Electrolytic cell is any device in which electrical energy is converted to chemical energy, or vice versa. Such a cell typically consists of two metallic or electronic conductors (electrodes) held apart from each other and in contact with an electrolyte (*q.v.*), usually a dissolved or fused ionic compound. Connection of the electrodes to a source of direct electric current renders one of them negatively charged and the other positively charged. Positive ions in the electrolyte migrate to the negative electrode (cathode) and there combine with one or more electrons, losing part or all of their charge and becoming new ions having lower charge or neutral atoms or molecules; at the same time, negative ions migrate to the positive electrode (anode) and transfer one or more electrons to it, also becoming new ions or neutral particles. The overall effect of the two processes is the transfer of electrons from the negative ions to the positive ions, a chemical reaction. An example is the electrolysis of sodium chloride (common salt), forming sodium metal and chlorine gas; the energy required to make the reaction proceed is supplied by the electric current. Other common applications of electrolysis include electrodeposition for refining or plating of metals and the production of caustic soda.

In the case of substances that generate energy, rather than consume it, when they react with each other, some or all of this energy can be converted to electricity if the reaction can be divided into an oxidation and a reduction that can be made to occur at separate electrodes. In the lead-acid storage battery, for example, lead dioxide, lead metal, and sulfuric acid react to form lead sulfate and water; the separate processes are the oxidation of lead to lead sulfate at one electrode and the reduction of lead dioxide to lead sulfate at the other while electric charge is transported through the electrolyte by the migration of hydrogen ions. These processes create a driving force (a voltage or electrical potential) that causes electricity to flow through an external circuit joining the two electrodes. Many other chemical combinations have been utilized in cells and batteries.

Other cells for generating electricity by means other than motion of a conductor in a magnetic field include solar cells, in which electron flow between semiconductors results from absorption of light, and fuel cells, in which a continuous supply of liquid or gaseous oxidizing agent, such as oxygen, removes electrons from the cathode as a reducing agent, such as hydrogen, supplies electrons to the anode.

Galvanic Cell

A galvanic cell or voltaic cell, named after Luigi Galvani or Alessandro Volta, respectively, is an electrochemical cell that derives electrical energy from spontaneous redox reactions taking place within the cell. It generally consists of two different metals immersed in an electrolyte, or of individual half-cells with different metals and their ions in solution connected by a salt bridge or separated by a porous membrane.

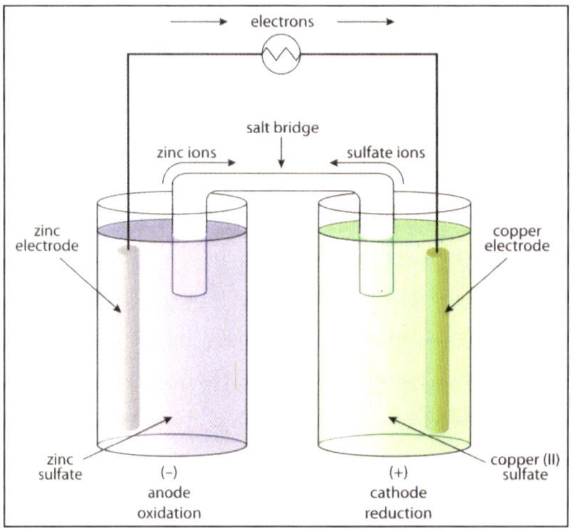

Schematic of Zn-Cu galvanic cell.

Volta was the inventor of the voltaic pile, the first electrical battery. In common usage, the word “battery” has come to include a single galvanic cell, but a battery properly consists of multiple cells.

In its simplest form, a half-cell consists of a solid metal (called an electrode) that is submerged in a solution; the solution contains cations (+) of the electrode metal and anions (–) to balance the charge of the cations. The full cell consists of two half-cells, usually connected by a semi-permeable membrane or by a salt bridge that prevents the ions of the more noble metal from plating out at the other electrode.

A specific example is the Daniell cell, with a zinc (Zn) half-cell containing a solution of $ZnSO_4$ (zinc sulfate) and a copper (Cu) half-cell containing a solution of $CuSO_4$ (copper sulfate). A salt bridge is used here to complete the electric circuit.

If an external electrical conductor connects the copper and zinc electrodes, zinc from the zinc electrode dissolves into the solution as Zn^{2+} ions (oxidation), releasing electrons that enter the external conductor. To compensate for the increased zinc ion concentration, via the salt bridge zinc ions leave and anions enter the zinc half-cell. In the copper half-cell, the copper ions plate onto the copper electrode (reduction), taking up electrons that leave the external conductor. Since the Cu^{2+} ions (cations) plate onto the copper electrode, the latter is called the cathode. Correspondingly the zinc electrode is the anode. The electrochemical reaction is:

$$Zn + Cu^{2+} \rightarrow Zn^{2+} + Cu$$

In addition, electrons flow through the external conductor, which is the primary application of the galvanic cell.

The emf of the cell is the difference of the half-cell potentials, a measure of the relative ease of dissolution of the two electrodes into the electrolyte. The emf depends on both the electrodes and on the electrolyte, an indication that the emf is chemical in nature.

Electrochemical Thermodynamics of Galvanic Cell Reactions

The electrochemical processes in a galvanic cell occur because reactants of high free energy (e.g. metallic Zn and hydrated Cu^{2+} in the Daniell cell) are converted to lower-energy products (metallic Cu and hydrated Zn^{2+} in this example). The difference in the lattice cohesive energies of the electrode metals is sometimes the dominant energetic driver of the reaction, specifically in the Daniell cell. Metallic Zn, Cd, Li, and Na, which are not stabilized by d-orbital bonding, have higher cohesive energies (i.e. they are more weakly bonded) than all transition metals, including Cu, and are therefore useful as high-energy anode metals.

The difference between the metals' ionization energies in water is the other energetic contribution that can drive the reaction in a galvanic cell; it is not important in the Daniell cell because the energies of hydrated Cu^{2+} and Zn^{2+} ions happen to be similar. Both atom transfer, e.g. of zinc from the metal electrode into the solution, and electron transfer, from metal atoms or to metal ions, play important roles in a galvanic cell. Concentration cells, whose electrodes and ions are made of the same metal and which are driven by an entropy increase and free-energy decrease as ion concentrations equalize, show that the electronegativity difference of the metals is not the driving force of electrochemical processes.

Galvanic cells and batteries are typically used as a source of electrical power. The energy derives from a high-cohesive-energy metal dissolving while to a lower-energy metal is deposited, and/or from high-energy metal ions plating out while lower-energy ions go into solution.

Quantitatively, the electrical energy produced by a galvanic cell is approximately equal to the standard free-energy difference of the reactants and products, denoted as $\Delta_r G^o$. In the Daniell cell, most of the electrical energy of $\Delta_r G^o$ = -213 kJ/mol can be attributed to the -207 kJ/mol difference between Zn and Cu lattice cohesive energies.

Half Reactions and Conventions

A half-cell contains a metal in two oxidation states. Inside an isolated half-cell, there is an oxidation-reduction (redox) reaction that is in chemical equilibrium, a condition written symbolically as follows (here, "M" represents a metal cation, an atom that has a charge imbalance due to the loss of "n" electrons):

$$M^{n+} \text{ (oxidized species) } + ne^- \rightleftharpoons M \text{ (reduced species)}$$

A galvanic cell consists of two half-cells, such that the electrode of one half-cell is composed of metal A, and the electrode of the other half-cell is composed of metal B; the redox reactions for the two separate half-cells are thus:

$$A^{n+} + ne^- \rightleftharpoons A$$

$$B^{m+} + me^- \rightleftharpoons B$$

The overall balanced reaction is:

$$m\,A + n\,B^{m+} \rightleftharpoons n\,B + m\,A^{n+}$$

In other words, the metal atoms of one half-cell are oxidized while the metal cations of the other half-cell are reduced. By separating the metals in two half-cells, their reaction can be controlled in a way that forces transfer of electrons through the external circuit where they can do useful work.

The electrodes are connected with a metal wire in order to conduct the electrons that participate in the reaction.

In one half-cell, dissolved metal-B cations combine with the free electrons that are available at the interface between the solution and the metal-B electrode; these cations are thereby neutralized, causing them to precipitate from solution as deposits on the metal-B electrode, a process known as plating.

This reduction reaction causes the free electrons throughout the metal-B electrode, the wire, and the metal-A electrode to be pulled into the metal-B electrode. Consequently, electrons are wrestled away from some of the atoms of the metal-A electrode, as though the metal-B cations were reacting directly with them; those metal-A atoms become cations that dissolve into the surrounding solution.

As this reaction continues, the half-cell with the metal-A electrode develops a positively charged solution (because the metal-A cations dissolve into it), while the other half-cell develops a negatively charged solution (because the metal-B cations precipitate out of it, leaving behind the anions); unabated, this imbalance in charge would stop the reaction. The solutions of the half-cells are connected by a salt bridge or a porous plate that allows ions to pass from one solution to the other, which balances the charges of the solutions and allows the reaction to continue.

The anode is the electrode where oxidation (loss of electrons) takes place (metal-A electrode); in a galvanic cell, it is the negative electrode, because when oxidation occurs, electrons are left behind on the electrode. These electrons then flow through the external circuit to the cathode (positive electrode) (while in electrolysis, an electric current drives electron flow in the opposite direction and the anode is the positive electrode).

The cathode is the electrode where reduction (gain of electrons) takes place (metal-B electrode); in a galvanic cell, it is the positive electrode, as ions get reduced by taking up electrons from the electrode and plate out (while in electrolysis, the cathode is the negative terminal and attracts positive ions from the solution). In both cases, the statement 'the cathode attracts cations' is true.

Galvanic cells, by their nature, produce direct current. The Weston cell has an anode composed of cadmium mercury amalgam, and a cathode composed of pure mercury. The electrolyte is a (saturated) solution of cadmium sulfate. The depolarizer is a paste of mercurous sulfate. When the electrolyte solution is saturated, the voltage of the cell is very reproducible; hence, in 1911, it was adopted as an international standard for voltage.

A battery is a set of galvanic cells that are connected together to form a single source of voltage. For instance, a typical 12V lead–acid battery has six galvanic cells connected in series with the anodes composed of lead and cathodes composed of lead dioxide, both immersed in sulfuric acid. Large battery rooms, for instance in a telephone exchange providing central office power to user's telephones, may have cells connected in both series and parallel.

Cell Voltage

The voltage (electromotive force E^o) produced by a galvanic cell can be estimated from the standard Gibbs free energy change in the electrochemical reaction according to:

$$E^o_{cell} = -\Delta_r G^o / (\nu_e F)$$

where ν_e is the number of electrons transferred in the balanced half reactions, and F is Faraday's constant. However, it can be determined more conveniently by the use of a standard potential table for the two half cells involved. The first step is to identify the two metals and their ions reacting in the cell. Then one looks up the standard electrode potential, E^o, in volts, for each of the two half reactions. The standard potential of the cell is equal to the more positive E^o value minus the more negative E^o value.

For example, the solutions are $CuSO_4$ and $ZnSO_4$. Each solution has a corresponding metal strip in it, and a salt bridge or porous disk connecting the two solutions and allowing SO_4^{2-} ions to flow freely between the copper and zinc solutions. To calculate the standard potential one looks up copper and zinc's half reactions and finds:

$$Cu^{2+} + 2\,e^{-} \rightleftharpoons Cu \quad E^{o} = +0.34\ V$$

$$Zn^{2+} + 2\,e^{-} \rightleftharpoons Zn \quad E^{o} = -0.76\ V$$

Thus the overall reaction is:

$$Cu^{2+} + Zn \rightleftharpoons Cu + Zn^{2+}$$

The standard potential for the reaction is then +0.34 V – (−0.76 V) = 1.10 V. The polarity of the cell is determined as follows. Zinc metal is more strongly reducing than copper metal because the standard (reduction) potential for zinc is more negative than that of copper. Thus, zinc metal will lose electrons to copper ions and develop a positive electrical charge. The equilibrium constant, *K*, for the cell is given by:

$$\ln K = \frac{\nu_e F E^{o}_{cell}}{RT}$$

where *F* is the Faraday constant, *R* is the gas constant and *T* is the temperature in kelvins. For the Daniell cell *K* is approximately equal to 1.5×10^{37}. Thus, at equilibrium, a few electrons are transferred, enough to cause the electrodes to be charged.

Actual half-cell potentials must be calculated by using the Nernst equation as the solutes are unlikely to be in their standard states,

$$E_{half\text{-}cell} = E^{o} - \frac{RT}{\nu_e F}\ln_e Q$$

where *Q* is the reaction quotient. When the charges of the ions in the reaction are equal, this simplifies to:

$$E_{half\text{-}cell} = E^{o} - 2.303\frac{RT}{\nu_e F}\log_{10}\left\{M^{n+}\right\}$$

where $\{M^{n+}\}$ is the activity of the metal ion in solution. In practice concentration in mol/L is used in place of activity. The metal electrode is in its standard state so by definition has unit activity. The potential of the whole cell is obtained as the difference between the potentials for the two half-cells, so it depends on the concentrations of both dissolved metal ions. If the concentrations are the same, $E_{cell} = E^{o}_{cell}$ and the Nernst equation is not needed under the conditions assumed here.

The value of $2.303R/F$ is 1.9845×10^{-4} V/K, so at 25 °C (298.15 K) the half-cell potential will change by only 0.05918 V/v_e if the concentration of a metal ion is increased or decreased by a factor of 10.

$$E_{half\text{-}cell} = E^o - \frac{0.05918\,V}{v_e}\log_{10}\left[M^{n+}\right]$$

These calculations are based on the assumption that all chemical reactions are in equilibrium. When a current flows in the circuit, equilibrium conditions are not achieved and the cell voltage will usually be reduced by various mechanisms, such as the development of overpotentials. Also, since chemical reactions occur when the cell is producing power, the electrolyte concentrations change and the cell voltage is reduced. A consequence of the temperature dependency of standard potentials is that the voltage produced by a galvanic cell is also temperature dependent.

Galvanic Corrosion

Galvanic corrosion is a process that degrades metals electrochemically. This corrosion occurs when two dissimilar metals are in contact with each other in the presence of an electrolyte, such as salt water, forming a galvanic cell with H_2 formation on the more noble metal. The resulting electrochemical potential then develops an electric current that electrolytically dissolves the less noble material. A concentration cell can be formed if the same metal is exposed to two different concentrations of electrolyte.

Cell Types

- Concentration cell
- Electrolytic cell
- Electrochemical cell
- Lemon battery
- Thermogalvanic cell

Primary Cell

A primary cell is a battery (a galvanic cell) that is designed to be used once and discarded, and not recharged with electricity and reused like a secondary cell (rechargeable battery). In general, the electrochemical reaction occurring in the cell is not reversible, rendering the cell unrechargeable. As a primary cell is used, chemical reactions in the battery use up the chemicals that generate the power; when they arc gone, the battery stops producing electricity and is useless. In contrast, in a secondary cell, the reaction can be reversed by running a current into the cell with a battery charger to recharge it, regenerating the chemical reactants. Primary cells are made in a range of standard sizes to power small household appliances such as flashlights and portable radios.

Primary batteries make up about 90% of the $50 billion battery market, but secondary batteries have been gaining market share. About 15 billion primary batteries are thrown away worldwide every year, virtually all ending up in landfills. Due to the toxic heavy metals and strong acids they contain, batteries are hazardous waste. Most municipalities classify them as such and require separate disposal. The energy needed to manufacture a battery is about 50 times greater than the energy it contains. Due to their high pollutant content compared to their small energy content, the primary battery is considered a wasteful, environmentally unfriendly technology. Due mainly to increasing sales of wireless devices and cordless tools which cannot be economically powered by primary batteries and come with integral rechargeable batteries, the secondary battery industry has high growth and has slowly been replacing the primary battery in high end products.

In the early twenty-first century, primary cells began losing market share to secondary cells, as relative costs declined for the latter. Flashlight power demands were reduced by the switch from incandescent bulbs to light-emitting diodes.

The remaining market experienced increased competition from private- or no-label versions. The market share of the two leading US manufacturers, Energizer and Duracell, declined to 37% in 2012. Along with Rayovac, these three are trying to move consumers from zinc-carbon to more expensive, long-lasting and safer alkaline batteries.

Western battery manufacturers shifted production offshore and no longer make zinc-carbon batteries in the United States.

China became the largest battery market, with demand projected to climb faster than anywhere else, and has also shifted to alkaline cells. In other developing countries disposable batteries must compete with cheap wind-up, wind-powered and rechargeable devices that have proliferated.

Comparison between Primary and Secondary Cells

Secondary cells (rechargeable batteries) are in general more economical to use than primary cells. Their initially higher cost and the purchase cost of a charging system can be spread out over many use cycles (between 100 and 1000 cycles); for example, in hand-held power tools, it would be very costly to replace a high-capacity primary battery pack every few hours of use.

Primary cells are not designed for recharging between manufacturing and use, thus have battery chemistry that has to have a much lower self-discharge rate than older types of secondary cells; but they have lost that advantage with the development of rechargeable secondary cells with very low self discharge rates like low self-discharge NiMH cells that hold enough charge for long enough to be sold as pre-charged.

Common types of secondary cells (namely NiMH and Li-ion) due to their much lower internal resistance do not suffer the large loss of capacity that alkaline, zinc–carbon and zinc chloride ("heavy duty" or "super heavy duty") do with high current draw.

Reserve batteries achieve very long storage time (on the order of 10 years or more) without loss of capacity, by physically separating the components of the battery and only assembling them at the time of use. Such constructions are expensive but are found in applications like munitions, which may be stored for years before use.

Recharging Primary Cells

Attempts to extend the life of primary cells, for example by recharging alkaline batteries, meet with variable results. Third-party devices are manufactured and represented as being able to recharge primary cells. The internal chemical reactions of a primary cell are not easily reversed by externally applied currents, so reactants do not entirely return to their initial state and location. Recharged primary cells will not have the life or performance of secondary cells. Manufacturers of primary cells usually warn against recharging.

Polarization

A major factor reducing the lifetime of primary cells is that they become *polarized* during use. This means that hydrogen accumulates at the cathode and reduces the effectiveness of the cell. To reduce the effects of polarization in commercial cells and to extend their lives, chemical depolarization is used; that is, an oxidizing agent is added to the cell, to oxidize the hydrogen to water. Manganese dioxide is used in the Leclanché cell and zinc–carbon cell, and nitric acid is used in the Bunsen cell and Grove cell.

Attempts have been made to make simple cells self-depolarizing by roughening the surface of the copper plate to facilitate the detachment of hydrogen bubbles with little success. Electrochemical depolarization exchanges the hydrogen for a metal, such as copper (e.g. Daniell cell), or silver (e.g. silver-oxide cell), so called.

Anode and Cathode

The battery terminal (electrode) that develops a positive voltage polarity (the carbon electrode in a dry cell) is called the cathode and the electrode with a negative polarity (zinc in a dry cell) is called the anode. This is the *reverse* of the terminology used in an electrolytic cell or thermionic vacuum tube. The reason is that the terms anode and cathode are defined by the direction of electric current, not by their voltage. The anode is the terminal through which conventional current (positive charge) enters the cell from the external circuit, while the cathode is the terminal through which conventional current leaves the cell and flows into the external circuit. Since a battery is a power source which provides the voltage which forces the current through the external circuit, the voltage on the cathode must be higher than the voltage on the anode, creating an electric field directed from cathode to anode, to force the positive charge out of the cathode through the resistance of the external circuit.

Inside the cell the anode is the electrode where chemical oxidation occurs, as it donates electrons which flow out of the anode into the external circuit. The cathode is the electrode where chemical reduction occurs, as it accepts electrons from the circuit through the cathode.

Outside the cell, different terminology is used. As the anode donates positive charge to the electrolyte (thus remaining with an excess of electrons that it will donate to the circuit), it becomes negatively charged and is therefore connected to the terminal marked "–" on the outside of the cell. The cathode, meanwhile, donates negative charge to the electrolyte, so it becomes positively charged (which allows it to accept electrons from the circuit) and is therefore connected to the terminal marked "+" on the outside of the cell.

Old textbooks sometimes contain different terminology that can cause confusion to modern readers. For example, a 1911 textbook by Ayrton and Mather describes the electrodes as the "positive plate" and "negative plate".

Fuel Cell

A fuel cell is an electrochemical cell that converts the chemical energy of a fuel (often hydrogen) and an oxidizing agent (often oxygen) into electricity through a pair of redox reactions. Fuel cells are different from most batteries in requiring a continuous source of fuel and oxygen (usually from air) to sustain the chemical reaction, whereas in a battery the chemical energy usually comes from metals and their ions or oxides that are commonly already present in the battery, except in flow batteries. Fuel cells can produce electricity continuously for as long as fuel and oxygen are supplied.

The first fuel cells were invented by Sir William Grove in 1838. The first commercial use of fuel cells came more than a century later following the invention of the hydrogen–oxygen fuel cell by Francis Thomas Bacon in 1932. The alkaline fuel cell, also known as the Bacon fuel cell after its inventor, has been used in NASA space programs since the mid-1960s to generate power for satellites and space capsules. Since then, fuel cells have been used in many other applications. Fuel cells are used for primary and backup power for commercial, industrial and residential buildings and in remote or inaccessible areas. They are also used to power fuel cell vehicles, including forklifts, automobiles, buses, boats, motorcycles and submarines.

There are many types of fuel cells, but they all consist of an anode, a cathode, and an electrolyte that allows ions, often positively charged hydrogen ions (protons), to move between the two sides of the fuel cell. At the anode a catalyst causes the fuel to undergo oxidation reactions that generate ions (often positively charged hydrogen ions) and electrons. The ions move from the anode to the cathode through the electrolyte. At the same time, electrons flow from the anode to the cathode through an external circuit, producing direct current electricity. At the cathode, another catalyst causes ions, electrons, and oxygen to react, forming water and possibly other products. Fuel cells are classified by the type of electrolyte they use and by the difference in startup time ranging from 1 second for proton exchange membrane fuel cells (PEM fuel cells, or PEMFC) to 10 minutes for solid oxide fuel cells (SOFC). A related technology is flow batteries, in which the fuel can be regenerated by recharging. Individual fuel cells produce relatively small electrical potentials, about 0.7 volts, so cells are "stacked", or placed in series, to create sufficient voltage to meet an application's requirements. In addition to electricity, fuel cells produce water, heat and, depending on the fuel source, very small amounts of nitrogen dioxide and other emissions. The energy efficiency of a fuel cell is generally between 40–60%; however, if waste heat is captured in a cogeneration scheme, efficiencies of up to 85% can be obtained.

The fuel cell market is growing, and in 2013 Pike Research estimated that the stationary fuel cell market will reach 50 GW by 2020.

Types of Fuel Cells

Fuel cells come in many varieties; however, they all work in the same general manner. They are made up of three adjacent segments: the anode, the electrolyte, and the cathode. Two chemical reactions occur at the interfaces of the three different segments. The net result of the two reactions is that fuel is consumed, water or carbon dioxide is created, and an electric current is created, which can be used to power electrical devices, normally referred to as the load.

At the anode a catalyst oxidizes the fuel, usually hydrogen, turning the fuel into a positively charged ion and a negatively charged electron. The electrolyte is a substance specifically designed so ions can pass through it, but the electrons cannot. The freed electrons travel through a wire creating the electric current. The ions travel through the electrolyte to the cathode. Once reaching the cathode, the ions are reunited with the electrons and the two react with a third chemical, usually oxygen, to create water or carbon dioxide.

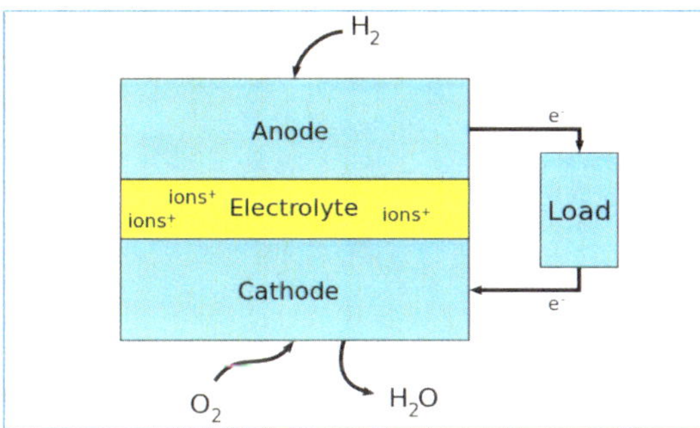

A block diagram of a fuel cell.

Design features in a fuel cell include:

- The electrolyte substance, which usually defines the *type* of fuel cell, and can be made from a number of substances like potassium hydroxide, salt carbonates, and phosphoric acid.
- The fuel that is used. The most common fuel is hydrogen.
- The anode catalyst, usually fine platinum powder, breaks down the fuel into electrons and ions.
- The cathode catalyst, often nickel, converts ions into waste chemicals, with water being the most common type of waste.
- Gas diffusion layers that are designed to resist oxidization.

A typical fuel cell produces a voltage from 0.6 V to 0.7 V at full rated load. Voltage decreases as current increases, due to several factors:

- Activation loss.
- Ohmic loss (voltage drop due to resistance of the cell components and interconnections).
- Mass transport loss (depletion of reactants at catalyst sites under high loads, causing rapid loss of voltage).

To deliver the desired amount of energy, the fuel cells can be combined in series to yield higher voltage, and in parallel to allow a higher current to be supplied. Such a design is called a *fuel cell stack*. The cell surface area can also be increased, to allow higher current from each cell. Within the stack, reactant gases must be distributed uniformly over each of the cells to maximize the power output.

Proton-exchange Membrane Fuel Cells (PEMFCs)

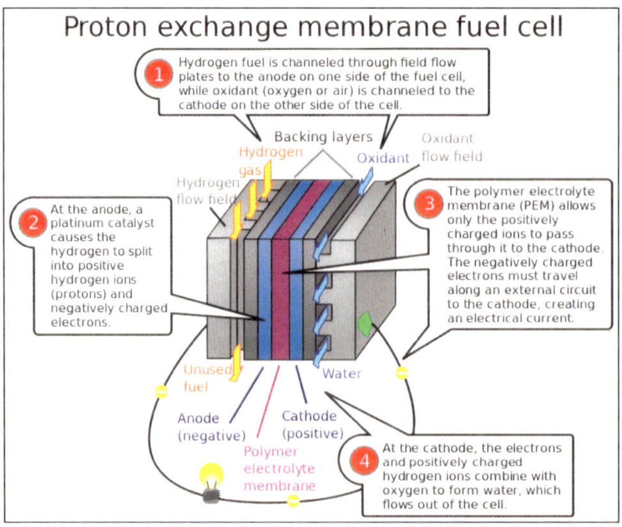

Construction of a high-temperature PEMFC: Bipolar plate as electrode with in-milled gas channel structure, fabricated from conductive composites (enhanced with graphite, carbon black, carbon fiber, and/or carbon nanotubes for more conductivity); Porous carbon papers; reactive layer, usually on the polymer membrane applied; polymer membrane.

Condensation of water produced by a PEMFC on the air channel wall. The gold wire around the cell ensures the collection of electric current.

In the archetypical hydrogen–oxide proton-exchange membrane fuel cell design, a proton-conducting polymer membrane (typically nafion) contains the electrolyte solution that separates the anode and cathode sides. This was called a "solid polymer electrolyte fuel cell" (SPEFC) in the early 1970s, before the proton exchange mechanism was well understood. (Notice that the synonyms "polymer electrolyte membrane" and "proton exchange mechanism" result in the same acronym.)

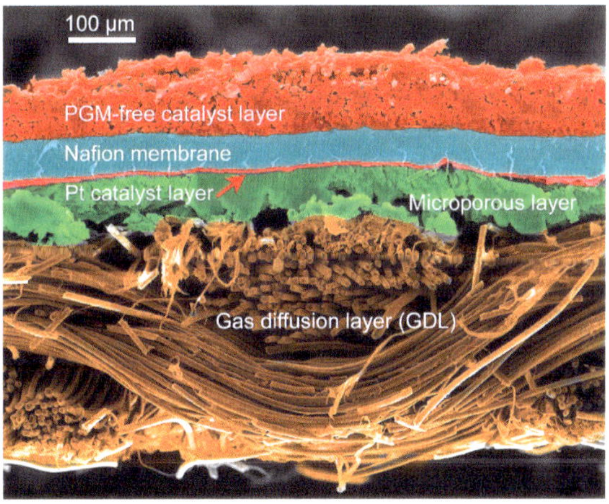

SEM micrograph of a PEMFC MEA cross-section with a non-precious metal catalyst cathode and Pt/C anode. False colors applied for clarity.

On the anode side, hydrogen diffuses to the anode catalyst where it later dissociates into protons and electrons. These protons often react with oxidants causing them to become what are commonly referred to as multi-facilitated proton membranes. The protons are conducted through the membrane to the cathode, but the electrons are forced to travel in an external circuit (supplying power) because the membrane is electrically insulating. On the cathode catalyst, oxygen molecules react with the electrons (which have traveled through the external circuit) and protons to form water.

In addition to this pure hydrogen type, there are hydrocarbon fuels for fuel cells, including diesel, methanol and chemical hydrides. The waste products with these types of fuel are carbon dioxide and water. When hydrogen is used, the CO_2 is released when methane from natural gas is combined with steam, in a process called steam methane reforming, to produce the hydrogen. This can take place in a different location to the fuel cell, potentially allowing the hydrogen fuel cell to be used indoors—for example, in fork lifts.

The different components of a PEMFC are:

1. Bipolar plates,
2. Electrodes,
3. Catalyst,
4. Membrane,
5. The necessary hardware such as current collectors and gaskets.

The materials used for different parts of the fuel cells differ by type. The bipolar plates may be made of different types of materials, such as, metal, coated metal, graphite, flexible graphite, C–C composite, carbon–polymer composites etc. The membrane electrode assembly (MEA) is referred as the heart of the PEMFC and is usually made of a proton exchange membrane sandwiched between two catalyst-coated carbon papers. Platinum and/or similar type of noble metals are usually used as the catalyst for PEMFC. The electrolyte could be a polymer membrane.

Proton Exchange Membrane Fuel Cell Design Issues

- Cost: In 2013, the Department of Energy estimated that 80-kW automotive fuel cell system costs of US$67 per kilowatt could be achieved, assuming volume production of 100,000 automotive units per year and US$55 per kilowatt could be achieved, assuming volume production of 500,000 units per year. Many companies are working on techniques to reduce cost in a variety of ways including reducing the amount of platinum needed in each individual cell. Ballard Power Systems has experimented with a catalyst enhanced with carbon silk, which allows a 30% reduction (1 mg/cm^2 to 0.7 mg/cm^2) in platinum usage without reduction in performance. Monash University, Melbourne uses PEDOT as a cathode. A 2011 published study doi: 10.1021/ja1112904 documented the first metal-free electrocatalyst using relatively inexpensive doped carbon nanotubes, which are less than 1% the cost of platinum and are of equal or superior performance.

- Water and air management (in PEMFCs): In this type of fuel cell, the membrane must be hydrated, requiring water to be evaporated at precisely the same rate that it is produced. If water is evaporated too quickly, the membrane dries, resistance across it increases, and eventually it will crack, creating a gas "short circuit" where hydrogen and oxygen combine directly, generating heat that will damage the fuel cell. If the water is evaporated too slowly, the electrodes will flood, preventing the reactants from reaching the catalyst and stopping the reaction. Methods to manage water in cells are being developed like electroosmotic pumps focusing on flow control. Just as in a combustion engine, a steady ratio between the reactant and oxygen is necessary to keep the fuel cell operating efficiently.

- Temperature management: The same temperature must be maintained throughout the cell in order to prevent destruction of the cell through thermal loading. This is particularly challenging as the $2H_2 + O_2 \rightarrow 2H_2O$ reaction is highly exothermic, so a large quantity of heat is generated within the fuel cell.
- Durability, service life, and special requirements for some type of cells: Stationary fuel cell applications typically require more than 40,000 hours of reliable operation at a temperature of −35 °C to 40 °C (−31 °F to 104 °F), while automotive fuel cells require a 5,000-hour lifespan (the equivalent of 240,000 km (150,000 mi)) under extreme temperatures. Current service life is 2,500 hours (about 75,000 miles). Automotive engines must also be able to start reliably at −30 °C (−22 °F) and have a high power-to-volume ratio (typically 2.5 kW per liter).
- Limited carbon monoxide tolerance of some (non-PEDOT) cathodes.

Phosphoric Acid Fuel Cell (PAFC)

Phosphoric acid fuel cells (PAFC) were first designed and introduced in 1961 by G. V. Elmore and H. A. Tanner. In these cells phosphoric acid is used as a non-conductive electrolyte to pass positive hydrogen ions from the anode to the cathode. These cells commonly work in temperatures of 150 to 200 degrees Celsius. This high temperature will cause heat and energy loss if the heat is not removed and used properly. This heat can be used to produce steam for air conditioning systems or any other thermal energy consuming system. Using this heat in cogeneration can enhance the efficiency of phosphoric acid fuel cells from 40–50% to about 80%. Phosphoric acid, the electrolyte used in PAFCs, is a non-conductive liquid acid which forces electrons to travel from anode to cathode through an external electrical circuit. Since the hydrogen ion production rate on the anode is small, platinum is used as catalyst to increase this ionization rate. A key disadvantage of these cells is the use of an acidic electrolyte. This increases the corrosion or oxidation of components exposed to phosphoric acid.

Solid Acid Fuel Cell (SAFC)

Solid acid fuel cells (SAFCs) are characterized by the use of a solid acid material as the electrolyte. At low temperatures, solid acids have an ordered molecular structure like most salts. At warmer temperatures (between 140 and 150 degrees Celsius for $CsHSO_4$), some solid acids undergo a phase transition to become highly disordered "superprotonic" structures, which increases conductivity by several orders of magnitude. The first proof-of-concept SAFCs were developed in 2000 using cesium hydrogen sulfate ($CsHSO_4$). Current SAFC systems use cesium dihydrogen phosphate (CsH_2PO_4) and have demonstrated lifetimes in the thousands of hours.

Alkaline Fuel Cell (AFC)

The alkaline fuel cell or hydrogen-oxygen fuel cell was designed and first demonstrated publicly by Francis Thomas Bacon in 1959. It was used as a primary source of electrical energy in the Apollo space program. The cell consists of two porous carbon electrodes impregnated with a suitable catalyst such as Pt, Ag, CoO, etc. The space between the two electrodes is filled with a concentrated solution of KOH or NaOH which serves as an electrolyte. H_2 gas and O_2 gas are bubbled into the electrolyte through the porous carbon electrodes. Thus the overall reaction involves the combination of hydrogen gas and oxygen gas to form water. The cell runs continuously until the reactant's supply is exhausted. This type of cell operates efficiently in the temperature range 343 K to 413 K and provides a potential of about 0.9 V. AAEMFC is a type of AFC which employs a solid polymer electrolyte instead of aqueous potassium hydroxide (KOH) and it is superior to aqueous AFC.

High-temperature Fuel Cells

Solid Oxide Fuel Cell

Solid oxide fuel cells (SOFCs) use a solid material, most commonly a ceramic material called yttria-stabilized zirconia (YSZ), as the electrolyte. Because SOFCs are made entirely of solid materials, they are not limited to the flat plane configuration of other types of fuel cells and are often designed as rolled tubes. They require high operating temperatures (800–1000 °C) and can be run on a variety of fuels including natural gas.

SOFCs are unique since in those, negatively charged oxygen ions travel from the cathode (positive side of the fuel cell) to the anode (negative side of the fuel cell) instead of positively charged hydrogen ions travelling from the anode to the cathode, as is the case in all other types of fuel cells. Oxygen gas is fed through the cathode, where it absorbs electrons to create oxygen ions. The oxygen ions then travel through the electrolyte to react with hydrogen gas at the anode. The reaction at the anode produces electricity and water as by-products. Carbon dioxide may also be a by-product depending on the fuel, but the carbon emissions from an SOFC system are less than those from a fossil fuel combustion plant. The chemical reactions for the SOFC system can be expressed as follows:

Anode Reaction: $2H_2 + 2O^{2-} \rightarrow 2H_2O + 4e^-$

Cathode Reaction: $O_2 + 4e^- \rightarrow 2O^{2-}$

Overall Cell Reaction: $2H_2 + O_2 \rightarrow 2H_2O$

SOFC systems can run on fuels other than pure hydrogen gas. However, since hydrogen is necessary for the reactions listed above, the fuel selected must contain hydrogen atoms. For the fuel cell to operate, the fuel must be converted into pure hydrogen gas. SOFCs are capable of internally reforming light hydrocarbons such as methane (natural gas), propane and butane. These fuel cells are at an early stage of development.

Challenges exist in SOFC systems due to their high operating temperatures. One such challenge is the potential for carbon dust to build up on the anode, which slows down the internal reforming process. Research to address this "carbon coking" issue at the University of Pennsylvania has shown that the use of copper-based cermet (heat-resistant materials made of ceramic and metal) can reduce coking and the loss of performance. Another disadvantage of SOFC systems is slow start-up time, making SOFCs less useful for mobile applications. Despite these disadvantages, a high operating temperature provides an advantage by removing the need for a precious metal catalyst like platinum, thereby reducing cost. Additionally, waste heat from SOFC systems may be captured and reused, increasing the theoretical overall efficiency to as high as 80%–85%.

The high operating temperature is largely due to the physical properties of the YSZ electrolyte. As temperature decreases, so does the ionic conductivity of YSZ. Therefore, to obtain optimum performance of the fuel cell, a high operating temperature is required. According to their website, Ceres Power, a UK SOFC fuel cell manufacturer, has developed a method of reducing the operating temperature of their SOFC system to 500–600 degrees Celsius. They replaced the commonly used YSZ electrolyte with a CGO (cerium gadolinium oxide) electrolyte. The lower operating temperature allows them to use stainless steel instead of ceramic as the cell substrate, which reduces cost and start-up time of the system.

MCFC

Molten carbonate fuel cells (MCFCs) require a high operating temperature, 650 °C (1,200 °F), similar to SOFCs. MCFCs use lithium potassium carbonate salt as an electrolyte, and this salt liquefies at high temperatures, allowing for the movement of charge within the cell – in this case, negative carbonate ions.

Like SOFCs, MCFCs are capable of converting fossil fuel to a hydrogen-rich gas in the anode, eliminating the need to produce hydrogen externally. The reforming process creates CO_2 emissions. MCFC-compatible fuels include natural gas, biogas and gas produced from coal. The hydrogen in the gas reacts with carbonate ions from the electrolyte to produce water, carbon dioxide, electrons and small amounts of other chemicals. The electrons travel through an external circuit creating electricity and return to the cathode. There, oxygen from the air and carbon dioxide recycled from the anode react with the electrons to form carbonate ions that replenish the electrolyte, completing the circuit. The chemical reactions for an MCFC system can be expressed as follows:

Anode Reaction: $CO_3^{2-} + H_2 \rightarrow H_2O + CO_2 + 2e^-$

Cathode Reaction: $CO_2 + ½O_2 + 2e^- \rightarrow CO_3^{2-}$

Overall Cell Reaction: $H_2 + ½O_2 \rightarrow H_2O$

As with SOFCs, MCFC disadvantages include slow start-up times because of their high operating temperature. This makes MCFC systems not suitable for mobile applications, and this technology will most likely be used for stationary fuel cell purposes. The main challenge of MCFC technology is the cells' short life span. The high-temperature and carbonate electrolyte lead to corrosion of the anode and cathode. These factors accelerate the degradation of MCFC components, decreasing the durability and cell life. Researchers are addressing this problem by exploring corrosion-resistant materials for components as well as fuel cell designs that may increase cell life without decreasing performance.

MCFCs hold several advantages over other fuel cell technologies, including their resistance to impurities. They are not prone to "carbon coking", which refers to carbon build-up on the anode that results in reduced performance by slowing down the internal fuel reforming process. Therefore, carbon-rich fuels like gases made from coal are compatible with the system. The Department of Energy claims that coal, itself, might even be a fuel option in the future, assuming the system can be made resistant to impurities such as sulfur and particulates that result from converting coal into hydrogen. MCFCs also have relatively high efficiencies. They can reach a fuel-to-electricity efficiency of 50%, considerably higher than the 37–42% efficiency of a phosphoric acid fuel cell plant. Efficiencies can be as high as 65% when the fuel cell is paired with a turbine, and 85% if heat is captured and used in a Combined Heat and Power (CHP) system.

FuelCell Energy, a Connecticut-based fuel cell manufacturer, develops and sells MCFC fuel cells. The company says that their MCFC products range from 300 kW to 2.8 MW systems that achieve 47% electrical efficiency and can utilize CHP technology to obtain higher overall efficiencies. One product, the DFC-ERG, is combined with a gas turbine and, according to the company, it achieves an electrical efficiency of 65%.

Electric Storage Fuel Cell

The electric storage fuel cell is a conventional battery chargeable by electric power input, using the conventional electro-chemical effect. However, the battery further includes hydrogen (and oxygen) inputs for alternatively charging the battery chemically.

Efficiency of Leading Fuel Cell Types

Theoretical Maximum Efficiency

The energy efficiency of a system or device that converts energy is measured by the ratio of the amount of useful energy put out by the system ("output energy") to the total amount of energy that is put in ("input energy") or by useful output energy as a percentage of the total input energy. In the case of fuel cells, useful output energy is measured in electrical energy produced by the system. Input energy is the energy stored in the fuel. According to the U.S. Department of Energy, fuel cells are generally between 40–60% energy efficient. This is higher than some other systems for energy generation. For example, the typical internal combustion engine of a car is about 25% energy efficient. In combined heat and power (CHP) systems, the heat produced by the fuel cell is captured and put to use, increasing the efficiency of the system to up to 85–90%.

The theoretical maximum efficiency of any type of power generation system is never reached in practice, and it does not consider other steps in power generation, such as production, transportation and storage of fuel and conversion of the electricity into mechanical power. However, this calculation allows the comparison of different types of power generation. The maximum theoretical energy efficiency of a fuel cell is 83%, operating at low power density and using pure hydrogen and oxygen as reactants (assuming no heat recapture) According to the World Energy Council, this compares with a maximum theoretical efficiency of 58% for internal combustion engines.

In Practice

In a fuel-cell vehicle the tank-to-wheel efficiency is greater than 45% at low loads and shows average values of about 36% when a driving cycle like the NEDC (New European Driving Cycle) is used as test procedure. The comparable NEDC value for a Diesel vehicle is 22%. In 2008 Honda released a demonstration fuel cell electric vehicle (the Honda FCX Clarity) with fuel stack claiming a 60% tank-to-wheel efficiency.

It is also important to take losses due to fuel production, transportation, and storage into account. Fuel cell vehicles running on compressed hydrogen may have a power-plant-to-wheel efficiency of 22% if the hydrogen is stored as high-pressure gas, and 17% if it is stored as liquid hydrogen. Fuel cells cannot store energy like a battery, except as hydrogen, but in some applications, such as stand-alone power plants based on discontinuous sources such as solar or wind power, they are combined with electrolyzers and storage systems to form an energy storage system. Most hydrogen is used for oil refining, chemicals and fertilizer production and therefore produced by steam methane reforming, which emits carbon dioxide. The overall efficiency (electricity to hydrogen and back to electricity) of such plants (known as *round-trip efficiency*), using pure hydrogen and pure oxygen can be "from 35 up to 50 percent", depending on gas density and other conditions. The electrolyzer/fuel cell system can store indefinite quantities of hydrogen, and is therefore suited for long-term storage.

Solid-oxide fuel cells produce heat from the recombination of the oxygen and hydrogen. The ceramic can run as hot as 800 degrees Celsius. This heat can be captured and used to heat water in a micro combined heat and power (m-CHP) application. When the heat is captured, total efficiency can reach 80–90% at the unit, but does not consider production and distribution losses. CHP units are being developed today for the European home market.

Jeremy P. Meyers wrote, "While fuel cells are efficient relative to combustion engines, they are not as efficient as batteries, due primarily to the inefficiency of the oxygen reduction reaction (and the oxygen evolution reaction, should the hydrogen be formed by electrolysis of water). They make the most sense for operation disconnected from the grid, or when fuel can be provided continuously. For applications that require frequent and relatively rapid start-ups ... where zero emissions are a requirement, as in enclosed spaces such as warehouses, and where hydrogen is considered an acceptable reactant, a [PEM fuel cell] is becoming an increasingly attractive choice [if exchanging batteries is inconvenient]". In 2013 military organizations were evaluating fuel cells to determine if they could significantly reduce the battery weight carried by soldiers.

Applications

Type 212 submarine with fuel cell propulsion of the German Navy in dry dock.

Power

Stationary fuel cells are used for commercial, industrial and residential primary and backup power generation. Fuel cells are very useful as power sources in remote locations, such as spacecraft, remote weather stations, large parks, communications centers, rural locations including research stations, and in certain military applications. A fuel cell system running on hydrogen can be compact and lightweight, and have no major moving parts. Because fuel cells have no moving parts and do not involve combustion, in ideal conditions they can achieve up to 99.9999% reliability. This equates to less than one minute of downtime in a six-year period.

Since fuel cell electrolyzer systems do not store fuel in themselves, but rather rely on external storage units, they can be successfully applied in large-scale energy storage, rural areas being one example. There are many different types of stationary fuel cells so efficiencies vary, but most are between 40% and 60% energy efficient. However, when the fuel cell's waste heat is used to heat a building in a cogeneration system this efficiency can increase to 85%. This is significantly more efficient than traditional coal power plants, which are only about one third energy efficient. Assuming production at scale, fuel cells could save 20–40% on energy costs when used in cogeneration systems. Fuel cells are also much cleaner than traditional power generation; a fuel cell power plant using natural gas as a hydrogen source would create less than one ounce of pollution (other than CO_2) for every 1,000 kW·h produced, compared to 25 pounds of pollutants generated by conventional combustion systems. Fuel Cells also produce 97% less nitrogen oxide emissions than conventional coal-fired power plants.

One such pilot program is operating on Stuart Island in Washington State. There the Stuart Island Energy Initiative has built a complete, closed-loop system: Solar panels power an electrolyzer, which makes hydrogen. The hydrogen is stored in a 500-U.S.-gallon (1,900 L) tank at 200 pounds per square inch (1,400 kPa), and runs a ReliOn fuel cell to provide full electric back-up to the off-the-grid residence. Another closed system loop was unveiled in late 2011 in Hempstead, NY.

Fuel cells can be used with low-quality gas from landfills or waste-water treatment plants to generate power and lower methane emissions. A 2.8 MW fuel cell plant in California is said to be the largest of the type.

Cogeneration

Combined heat and power (CHP) fuel cell systems, including Micro combined heat and power (MicroCHP) systems are used to generate both electricity and heat for homes, office building and factories. The system generates constant electric power (selling excess power back to the grid when it is not consumed), and at the same time produces hot air and water from the waste heat. As the result CHP systems have the potential to save primary energy as they can make use of waste heat which is generally rejected by thermal energy conversion systems. A typical capacity range of home fuel cell is 1–3 kW_{el} / 4–8 kW_{th}. CHP systems linked to absorption chillers use their waste heat for refrigeration.

The waste heat from fuel cells can be diverted during the summer directly into the ground providing further cooling while the waste heat during winter can be pumped directly into the building.

Co-generation systems can reach 85% efficiency (40–60% electric + remainder as thermal). Phosphoric-acid fuel cells (PAFC) comprise the largest segment of existing CHP products worldwide and can provide combined efficiencies close to 90%. Molten

Carbonate (MCFC) and Solid Oxide Fuel Cells (SOFC) are also used for combined heat and power generation and have electrical energy efficiences around 60%. Disadvantages of co-generation systems include slow ramping up and down rates, high cost and short lifetime. Also their need to have a hot water storage tank to smooth out the thermal heat production was a serious disadvantage in the domestic market place where space in domestic properties is at a great premium.

Delta-ee consultants stated in 2013 that with 64% of global sales the fuel cell micro-combined heat and power passed the conventional systems in sales in 2012. The Japanese ENE FARM project will pass 100,000 FC mCHP systems in 2014, 34.213 PEMFC and 2.224 SOFC were installed in the period 2012-2014, 30,000 units on LNG and 6,000 on LPG.

Fuel Cell Electric Vehicles (FCEVs)

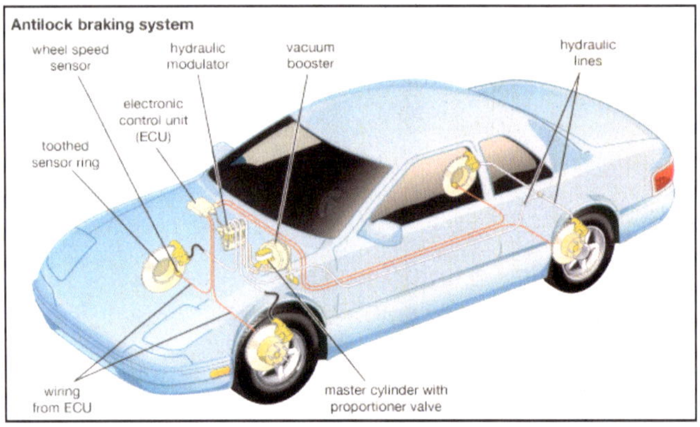

Configuration of components in a fuel cell car.

Toyota Mirai.

Automobiles

As of 2017, about 6500 FCEVs have been leased or sold worldwide. Three fuel cell electric vehicles have been introduced for commercial lease and sale: the Honda Clarity, Toyota Mirai and the Hyundai ix35 FCEV. Additional demonstration models include the Honda FCX Clarity, and Mercedes-Benz F-Cell. As of June 2011 demonstration FCEVs had driven more than 4,800,000 km (3,000,000 mi), with more than 27,000 refuelings. Fuel cell electric vehicles feature an average range of 314 miles between refuelings. They can be refueled in less than 5 minutes. The U.S. Department of Energy's Fuel Cell Technology Program states that, as of 2011, fuel cells achieved 53–59% efficiency at one-quarter power and 42–53% vehicle efficiency at full power, and a durability of over 120,000 km (75,000 mi) with less than 10% degradation. In a Well-to-Wheels simulation analysis that "did not address the economics and market constraints, General Motors and its partners estimated that per mile traveled, a fuel cell electric vehicle running on compressed gaseous hydrogen produced from natural gas could use about 40% less energy and emit 45% less greenhouse gasses than an internal combustion vehicle. A lead engineer from the Department of Energy whose team is testing fuel cell cars said in 2011 that the potential appeal is that "these are full-function vehicles with no limitations on range or refueling rate so they are a direct replacement for any vehicle. For instance, if you drive a full sized SUV and pull a boat up into the mountains, you can do that with this technology and you can't with current battery-only vehicles, which are more geared toward city driving."

In 2015, Toyota introduced its first fuel cell vehicle, the Mirai, at a price of $57,000. Hyundai introduced the limited production Hyundai ix35 FCEV under a lease agreement. In 2016, Honda started leasing the Honda Clarity Fuel Cell.

Buses

Toyota FCHV-BUS at the Expo.

As of August 2011, there were about 100 fuel cell buses running around the world, including in Whistler, Canada; San Francisco, United States; Hamburg, Germany; Shanghai, China; London, England; and São Paulo, Brazil. Most of these were manufactured by UTC Power, Toyota, Ballard, Hydrogenics, and Proton Motor. UTC buses had driven more than 970,000 km (600,000 mi) by 2011. Fuel cell buses have from 39% to 141% higher fuel economy than diesel buses and natural gas buses.

As of 2019, the NREL is evaluating several current and planned fuel cell bus projects in the U.S.

Forklifts

A fuel cell forklift (also called a fuel cell lift truck) is a fuel cell-powered industrial forklift truck used to lift and transport materials. In 2013 there were over 4,000 fuel cell forklifts used in material handling in the US, of which only 500 received funding from DOE. The global market is 1 million fork lifts per year. Fuel cell fleets are operated by various companies, including Sysco Foods, FedEx Freight, GENCO (at Wegmans, Coca-Cola, Kimberly Clark, and Whole Foods), and H-E-B Grocers. Europe demonstrated 30 fuel cell forklifts with Hylift and extended it with HyLIFT-EUROPE to 200 units, with other projects in France and Austria. Pike Research stated in 2011 that fuel cell-powered forklifts will be the largest driver of hydrogen fuel demand by 2020.

Most companies in Europe and the US do not use petroleum-powered forklifts, as these vehicles work indoors where emissions must be controlled and instead use electric forklifts. Fuel cell-powered forklifts can provide benefits over battery-powered forklifts as they can work for a full 8-hour shift on a single tank of hydrogen and can be refueled in 3 minutes. Fuel cell-powered forklifts can be used in refrigerated warehouses, as their performance is not degraded by lower temperatures. The FC units are often designed as drop-in replacements.

Motorcycles and Bicycles

In 2005 a British manufacturer of hydrogen-powered fuel cells, Intelligent Energy (IE), produced the first working hydrogen-run motorcycle called the ENV (Emission Neutral Vehicle). The motorcycle holds enough fuel to run for four hours, and to travel 160 km (100 mi) in an urban area, at a top speed of 80 km/h (50 mph). In 2004 Honda developed a fuel-cell motorcycle that utilized the Honda FC Stack.

Other examples of motorbikes and bicycles that use hydrogen fuel cells include the Taiwanese company APFCT's scooter using the fueling system from Italy's Acta SpA and the Suzuki Burgman scooter with an IE fuel cell that received EU Whole Vehicle Type Approval in 2011. Suzuki Motor Corp. and IE have announced a joint venture to accelerate the commercialization of zero-emission vehicles.

Airplanes

In 2003, the world's first propeller-driven airplane to be powered entirely by a fuel cell was flown. The fuel cell was a stack design that allowed the fuel cell to be integrated with the plane's aerodynamic surfaces. Fuel cell-powered unmanned aerial vehicles (UAV) include a Horizon fuel cell UAV that set the record distance flown for a small UAV in 2007. Boeing researchers and industry partners throughout Europe conducted experimental flight tests in February 2008 of a manned airplane powered only by a fuel cell and lightweight batteries. The fuel cell demonstrator airplane, as it was called, used a proton exchange membrane (PEM) fuel cell/lithium-ion battery hybrid system to power an electric motor, which was coupled to a conventional propeller.

In 2009 the Naval Research Laboratory's (NRL's) Ion Tiger utilized a hydrogen-powered fuel cell and flew for 23 hours and 17 minutes. Fuel cells are also being tested and considered to provide auxiliary power in aircraft, replacing fossil fuel generators that were previously used to start the engines and power on board electrical needs, while reducing carbon emissions.

In 2016 a Raptor E1 drone made a successful test flight using a fuel cell that was lighter than the lithium-ion battery it replaced. The flight lasted 10 minutes at an altitude of 80 metres (260 ft), although the fuel cell reportedly had enough fuel to fly for two hours. The fuel was contained in approximately 100 solid 1 square centimetre (0.16 sq in) pellets composed of a proprietary chemical within an unpressurized cartridge. The pellets are physically robust and operate at temperatures as warm as 50 °C (122 °F). The cell was from Arcola Energy.

Boats

The world's first certified fuel cell boat (HYDRA).

The world's first fuel-cell boat HYDRA used an AFC system with 6.5 kW net output. Iceland has committed to converting its vast fishing fleet to use fuel cells to provide auxiliary power by 2015 and, eventually, to provide primary power in its boats. Amsterdam recently introduced its first fuel cell-powered boat that ferries people around the city's canals.

Submarines

The Type 212 submarines of the German and Italian navies use fuel cells to remain submerged for weeks without the need to surface.

The U212A is a non-nuclear submarine developed by German naval shipyard Howaldtswerke Deutsche Werft. The system consists of nine PEM fuel cells, providing between

30 kW and 50 kW each. The ship is silent, giving it an advantage in the detection of other submarines. A naval paper has theorized about the possibility of a nuclear-fuel cell hybrid whereby the fuel cell is used when silent operations are required and then replenished from the Nuclear reactor (and water).

Portable Power Systems

Portable fuel cell systems are generally classified as weighing under 10 kg and providing power of less than 5 kW. The potential market size for smaller fuel cells is quite large with an up to 40% per annum potential growth rate and a market size of around $10 billion, leading a great deal of research to be devoted to the development of portable power cells. Within this market two groups have been identified. The first is the microfuel cell market, in the 1-50 W range for power smaller electronic devices. The second is the 1-5 kW range of generators for larger scale power generation (e.g. military outposts, remote oil fields).

Microfuel cells are primarily aimed at penetrating the market for phones and laptops. This can be primarily attributed to the advantageous energy density provided by fuel cells over a lithium-ion battery, for the entire system. For a battery, this system includes the charger as well as the battery itself. For the fuel cell this system would include the cell, the necessary fuel and peripheral attachments. Taking the full system into consideration, fuel cells have been shown to provide 530Wh/kg compared to 44 Wh/kg for lithium ion batteries. However, while the weight of fuel cell systems offer a distinct advantage the current costs are not in their favor. while a battery system will generally cost around $1.20 per Wh, fuel cell systems cost around $5 per Wh, putting them at a significant disadvantage.

As power demands for cell phones increase, fuel cells could become much more attractive options for larger power generation. The demand for longer on time on phones and computers is something often demanded by consumers so fuel cells could start to make strides into laptop and cell phone markets. The price will continue to go down as developments in fuel cells continues to accelerate. Current strategies for improving micro fuelcells is through the use of carbon nanotubes. It was shown by Girishkumar et al. that depositing nanotubes on electrode surfaces allows for substantially greater surface area increasing the oxygen reduction rate.

Fuel cells for use in larger scale operations also show much promise. Portable power systems that use fuel cells can be used in the leisure sector (i.e. RVs, cabins, marine), the industrial sector (i.e. power for remote locations including gas/oil wellsites, communication towers, security, weather stations), and in the military sector. SFC Energy is a German manufacturer of direct methanol fuel cells for a variety of portable power systems. Ensol Systems Inc. is an integrator of portable power systems, using the SFC Energy DMFC. The key advantage of fuel cells in this market is the great power generation per weight. While fuel cells can be expensive, for remote locations that require dependable energy fuel cells hold great power. For a 72-h excursion the comparison in weight is substantial, with a fuel cell only weighing 15 pounds compared to 29 pounds of batteries needed for the same energy.

Other Applications

- Providing power for base stations or cell sites.
- Distributed generation.
- Emergency power systems are a type of fuel cell system, which may include lighting, generators and other apparatus, to provide backup resources in a crisis or when regular systems fail. They find uses in a wide variety of settings from residential homes to hospitals, scientific laboratories, data centers.
- Telecommunication equipment and modern naval ships.
- An uninterrupted power supply (*UPS*) provides emergency power and, depending on the topology, provide line regulation as well to connected equipment by supplying power from a separate source when utility power is not available. Unlike a standby generator, it can provide instant protection from a momentary power interruption.
- Base load power plants.
- Solar Hydrogen Fuel Cell Water Heating.
- Hybrid vehicles, pairing the fuel cell with either an ICE or a battery.
- Notebook computers for applications where AC charging may not be readily available.
- Portable charging docks for small electronics (e.g. a belt clip that charges a cell phone or PDA).
- Smartphones, laptops and tablets.
- Small heating appliances.
- Food preservation, achieved by exhausting the oxygen and automatically maintaining oxygen exhaustion in a shipping container, containing, for example, fresh fish.
- Breathalyzers, where the amount of voltage generated by a fuel cell is used to determine the concentration of fuel (alcohol) in the sample.
- Carbon monoxide detector, electrochemical sensor.
- UAV such as Lockheed Martin Skunk Works Stalker, an electric UAV powered by solid oxide fuel cell with up to 8 hours of endurance compared 2 with traditional battery.

Fueling Stations

Hydrogen fueling station.

In 2013, *The New York Times* reported that there were "10 hydrogen stations available to the public in the entire United States: one in Columbia, S.C., eight in Southern California and the one in Emeryville". As of December 2016, there were 31 publicly accessible hydrogen refueling stations in the US, 28 of which were located in California.

A public hydrogen refueling station in Iceland operated from 2003 to 2007. It served three buses in the public transport net of Reykjavík. The station produced its own hydrogen with an electrolyzing unit. The 14 stations in Germany were planned to be expanded to 50 by 2015 through its public–private partnership Now GMBH.

By May 2017, there were 91 hydrogen fueling stations in Japan. As of 2016, Norway planned to build a network of hydrogen stations between the major cities, starting in 2017.

Half Cell

A half cell is one of the two electrodes of an electrochemical cell. An electrochemical cell comprises two half cells, where every half cell contains an electrode and an electrolyte. A salt bridge or direct contact is needed to connect two half cells. Without a connection between two half cells, no reaction will take place.

In a Zn-Cu battery, when two half cells are connected, the following reaction takes place:

Reduction reaction takes place at the cathode:

Cu^{2+} + 2 e --> Cu Cu^{2+} is the oxidizing agent and Cu the reducing agent.

Oxidation reaction takes place at the anode:

Zn --> Zn^{2+} + 2 e^- Zn is the reducing agent, and Zn^{2+} the oxidizing agent.

The electrons from the Zn will flow from the anode to the surface of the cathode to reduce copper (Cu^{2+}) ion and produce Cu. The oxidation-reduction takes place in a manner that the equilibrium of charges or electrons is established in an overall reaction. The overall reaction therefore involves the reduction of Cu^{2+} ions by zinc metal, as follows:

Zn(s)	+	Cu^{2+}(aq)	→	Zn^{2+}(aq)	+	Cu(s)

Since an oxidation reaction takes place at the anode and a reduction reaction takes place at the cathode of an electrochemical cell, the Zn/Zn^{2+} half cell becomes the anode, and the Cu^{2+}/Cu half cell becomes the cathode. The electrode potential developed by the overall cell reaction can be calculated through adding the potential developed both at the anode and cathode, i.e. at each half cell. Half-cell potential is very important in measuring the corrosion severity of a material.

References

- Electrochemical-reaction, science: britannica.com, Retrieved 29 June, 2019
- Danaher, kevin; biggs, shannon; mark, jason (2016). Building the green economy: success stories from the grassroots. Routledge. P. 199. Isbn 978-1317262923.
- Electrolytic-cell, technology: britannica.com, Retrieved 30 April, 2019
- Schmidt-rohr, k. (2018). "how batteries store and release energy: explaining basic electrochemistry" "j. Chem. Educ." 95: 1801-1810 , Retrieved 29 March, 2019
- Electrochem, chemical: hyperphysics.phy-astr.gsu.edu, Retrieved 2 March, 2019
- Hill, marquita k. (2004). Understanding environmental pollution: a primer. Cambridge university press. P. 274. Isbn 0521527260.
- Schmidt-rohr, k. (2015). "why combustions are always exothermic, yielding about 418 kj per mole of o2", j. Chem. Educ. 92: 2094-2099. Https://doi.org/10.1021/acs.jchemed.5b00333

Diverse Aspects of Chemical Engineering

5

- **Chemical Reactor**
- **Chemical Plant**
- **Transport Phenomena**
- **Chemical Process**
- **Unit Operation**
- **Process Design**
- **Separation Process**

Some of the fundamental aspects of chemical engineering include separation processes, unit operations, chemical kinetics, process design, etc. The topics elaborated in this chapter will help in gaining a better perspective about these processes of chemical engineering.

Chemical Reactor

The reactors, in which chemicals are made in industry, vary in size from a few cm^3 to the vast structures. For example, kilns that produce lime from limestone may be over 25 metres high and hold, at any one time, well over 400 tonnes of materials.

The design of the reactor is determined by many factors but of particular importance are the thermodynamics and kinetics of the chemical reactions being carried out. The two main types of reactor are termed *batch* and *continuous*.

Batch Reactors

Batch reactors are used for most of the reactions carried out in a laboratory. The reactants are placed in a test-tube, flask or beaker. They are mixed together, often heated for the reaction to take place and are then cooled. The products are poured out and, if necessary, purified.

This procedure is also carried out in industry, the key difference being one of size of reactor and the quantities of reactants.

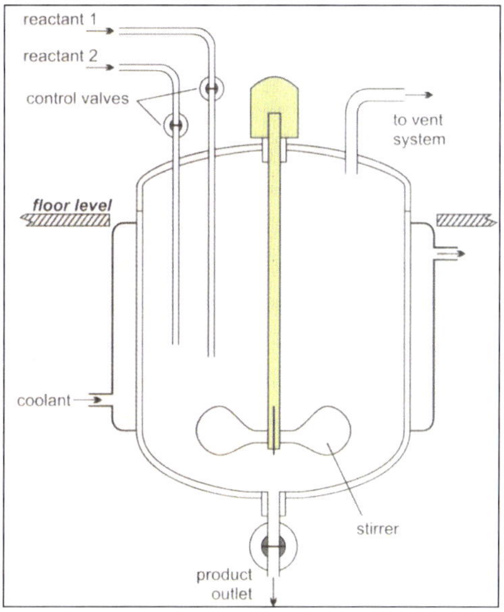

Illustrating a batch reactor.

Following reaction, the reactor is cleaned ready for another batch of reactants to be added.

Batch reactors are usually used when a company wants to produce a range of products involving different reactants and reactor conditions. They can then use the same equipment for these reactions.

Examples of processes that use batch reactors include the manufacture of colorants and margarine.

Colorants being produced in a batch reactor. The top of the reactor is at floor level and the rest of the reactor is suspended below it.

Continuous Reactors

An alternative to a batch process is to feed the reactants continuously into the reactor at one point, allow the reaction to take place and withdraw the products at another point. There must be an equal flow rate of reactants and products. While *continuous reactors* are rarely used in the laboratory, a water-softener can beregarded as an example of a continuous process. Hard water from the mains is passed through a tube containing an ion-exchange resin. Reaction occurs down the tube and soft water pours out at the exit.

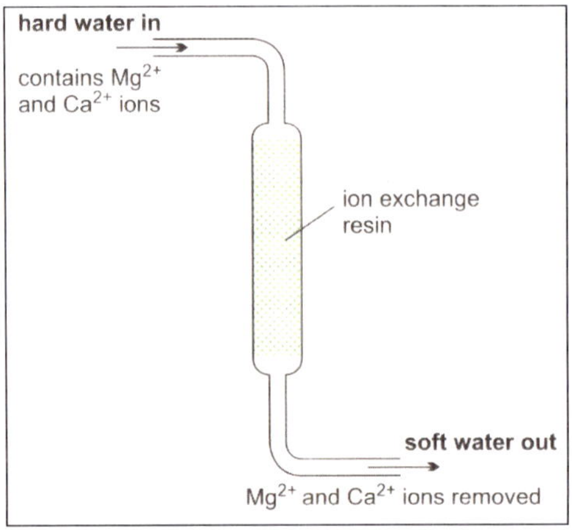

Illustrating a continuous reactor.

Continuous reactors are normally installed when large quantities of a chemical are being produced. It is important that the reactor can operate for several months without a shutdown.

The residence time in the reactor is controlled by the feed rate of reactants to the reactor. For example, if a reactor has a volume of 20 m^3 and the feed rate of reactants is 40 $m^3\,h^{-1}$ the residence time is 20 m^3 /40 $m^3\,h^{-1}$ = 0.5 h. It is simple to control accurately the flow rate of reactants. The volume is fixed and therefore the residence time in the reactor is also well controlled. The product tends to be of a more consistent quality from a continuous reactor because the reaction parameters (e.g. residence time, temperature and pressure) are better controlled than in batch operations. They also produce less waste and require much lower storage of both raw materials and products resulting in a more efficient operation. Capital costs per tonne of product produced are consequently lower. The main disadvantage is their lack of flexibility as once the reactor has been built it is only in rare cases that it can be used to perform a different chemical reaction.

Types of Continuous Reactors

Industry uses several types of continuous reactors.

(a) Tubular reactors: In a tubular reactor, fluids (gases and/or liquids) flow through it at high velocities. As the reactants flow, for example along a heated pipe, they are converted to products. At these high velocities, the products are unable to diffuse back and there is little or no back mixing. The conditions are referred to as plug flow. This reduces the occurrence of side reactions and increases the yield of the desired product.

With a constant flow rate, the conditions at any one point remain constant with time and changes in time of the reaction are measured in terms of the position along the length of the tube.

The reaction rate is faster at the pipe inlet because the concentration of reactants is at its highest and the reaction rate reduces as the reactants flow through the pipe due to the decrease in concentration of the reactant.

A tubular reactor used in the production of methyl 2-methylpropenoate. The reactor is heated by high pressure steam which has a temperature of 470 K and is fed into the reactor at point 1 and leaves the reactor at point 2. The reactants flow through the tubes.

Tubular reactors are used, for example, in the steam cracking of ethane, propane and butane and naphtha to produce alkenes.

(b) Fixed bed reactors: A heterogeneous catalyst is used frequently in industry where gases flow through a solid catalyst (which is often in the form of small pellets to increase the surface area). It is often described as a fixed bed of catalyst.

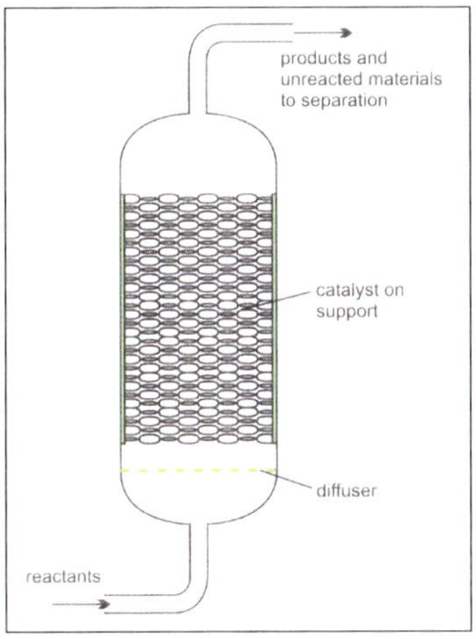

Illustrating a fixed bed reactor.

Among the examples of their use are the manufacture of sulfuric acid (the Contact Process, with vanadium(V) oxide as catalyst), the manufacure of nitric acid and the manufacture of ammonia (the Haber Process, with iron as the catalyst).

A further example of a fixed bed reactor is in catalytic reforming of naphtha to produce branched chain alkanes, cycloalkanes and aromatic hydrocarbons using usually platinum or a platinum-rhenium alloy on an alumina support.

(c) Fluid bed reactors: A fluid bed reactor is sometimes used whereby the catalyst particles, which are very fine, sit on a distributor plate. When the gaseous reactants pass through the distributor plate, the particles are carried with the gases forming a fluid. This ensures very good mixing of the reactants with the catalyst, with very high contact between the gaseous molecules and the catalyst and a good heat transfer. This results in a rapid reaction and a uniform mixture, reducing the variability of the process conditions.

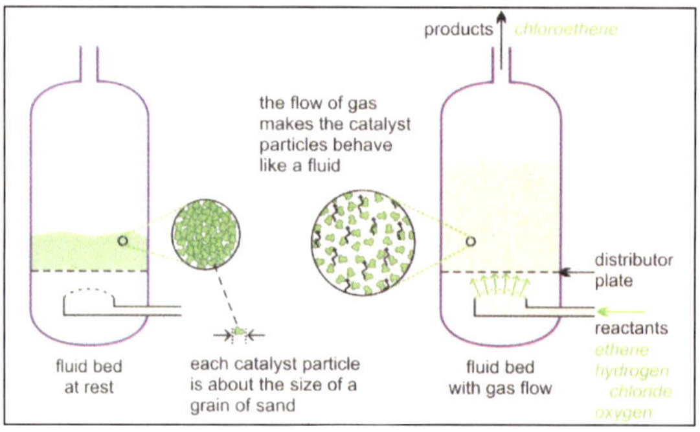

A diagram to illustrate a fluid bed reactor. On the left hand side, the particles are at rest. On the right hand side, the particles are now acting as a fluid, as the gaseous reactants pass through the solid.

One example of the use of fluid bed reactors is in the oxychlorination of ethene to chloroethene (vinyl chloride), the feedstock for the polymer poly(chloroethene) (PVC). The catalyst is copper(II) chloride and potassium chloride deposited on the surface of alumina. This support is so fine, it acts as a fluid when gases pass through it.

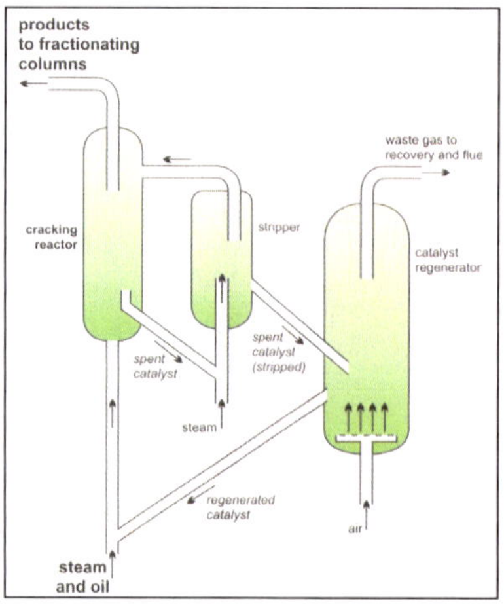

A catalytic cracker as used to produce alkenes from gas oil.

Another example is the catalytic cracking of gas oil to produce alkenes (ethene and propene) and petrol with a high octane rating.

A silica-alumina catalyst is used but fine particles of carbon are rapidly deposited on its surface and the catalyst quickly becomes ineffective. The catalyst, still in the form of a fluid, is regenerated as it passes into a second vessel where it is heated strongly in air (sometimes with added oxygen) burn off the carbon and then returned into the reaction zone and mixed with gas oil.

Part of the catalytic cracker at Fawley. 1 Fractionating column to remove and recover the butanes 2 The catalyst regenerator 3 Fractionating column to remove and recover ethane.

These reactors are larger than fixed bed reactors and are more expensive to construct. However, it is easier to control the conditions and the process is more efficient.

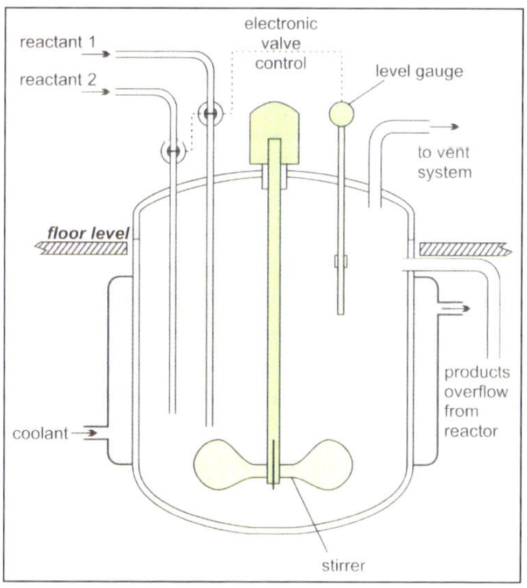

A line diagram illustrating a continuous stirred tank reactor.

(d) Continuous stirred tank reactors, CSTR: In a CSTR, one or more reactants, for example in solution or as a slurry, are introduced into a reactor equipped with an impeller (stirrer) and the products are removed continuously. The impeller stirs the reagents vigorously to ensure good mixing so that there is a uniform composition throughout. The composition at the outlet is the same as in the bulk in the reactor. These are exactly the opposite conditions to those in a tubular flow reactor where there is virtually no mixing of the reactants and the products.

A steady state must be reached where the flow rate into the reactor equals the flow rate out, for otherwise the tank would empty or overflow. The residence time is calculated by dividing the volume of the tank by the average volumetric flow rate. For example, if the flow of reactants is 10 $m^3 h^{-1}$ and the tank volume is 1 m^3, the residence time is 1/10 h, i.e. 6 minutes.

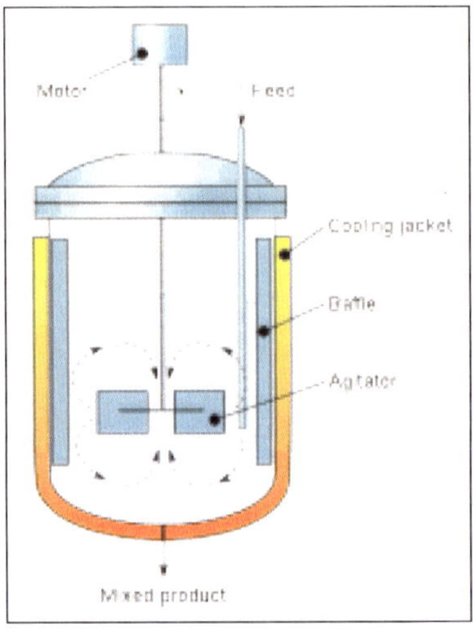

A CSTR reactor, used to make poly(ethene).

A CSTR reactor is used, for example in the production of the amide intermediate formed in the process to produce methyl 2-methylpropenoate. Sulfuric acid and 2-hydroxy-2-methylpropanonitrile are fed into the tank at a temperature of 400 K. The heat generated by the reaction is removed by cooling water fed through coils and the residence time is about 15 minutes.

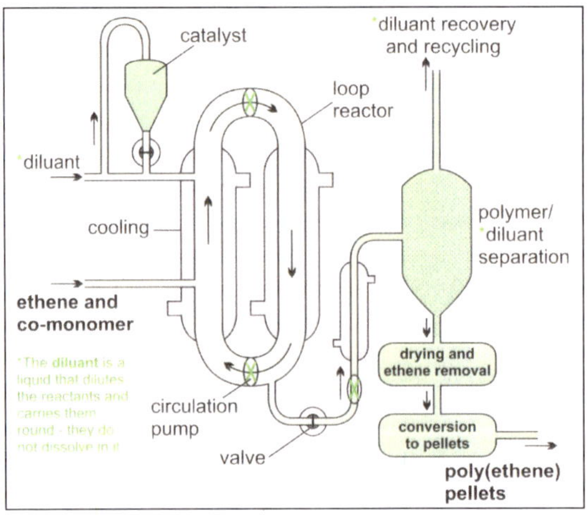

A line diagram showing a loop reactor used in the production of poly(ethene).

A variation of the CSTR is the loop reactor which is relatively simple and cheap to construct. In the diagram only one loop is shown. However, the residence time in the reactor is adjusted by altering the length or number of the loops in the reactor.

Loop reactors are used, for example, in the manufacture of poly(ethene) and the manufacture of poly(propene). Ethene (or propene) and the catalyst are mixed, under pressure, with a diluent, usually a hydrocarbon. A slurry is produced which is heated and circulated around the loops. Particles of the polymer gather at the bottom of one of the loop legs and, with some hydrocarbon diluent, are continuously released from the system. The diluent evaporates, leaving the solid polymer, and is then cooled to reform a liquid and passed back into the loop system, thus recirculating the hydrocarbon.

Heat Exchangers

Most chemical reactions are faster at higher temperatures and heat exchangers are frequently used to provide the heat necessary to increase the temperature of the reaction.

A common heat exchanger is the shell and tube type where one part of the process flows through a tube and the other part around the shell.

A good example where heat exchange is important is in the manufacture of sulfur trioxide from sulfur dioxide in the Contact Process where the excess heat is used to warm incoming gases.

The heat from the reaction is transferred to incoming gases across the tube wall and the rate of heat transfer is proportional to:

1. The temperature difference between the hot gases and the incoming gases.
2. The total surface area of the tubes.

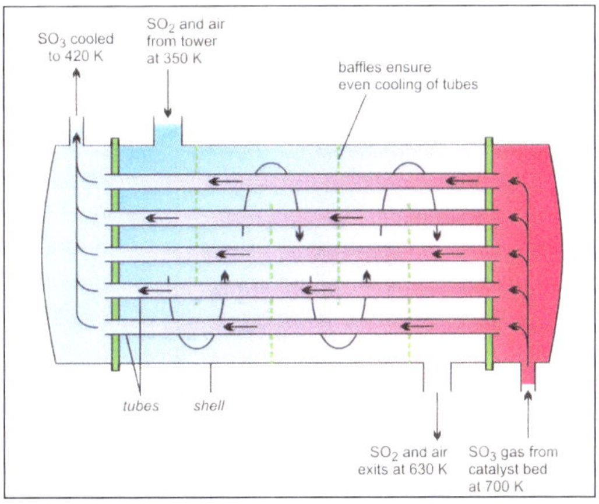

Illustrating a heat exchanger used in the manufacture of sulfur trioxide.

Thus the rate of heat transfer required will determine the size of the exchanger but when a chemical reaction also occurs in the exchanger (as in the case of tubular reactors), it is important to take into account the residence time of the materials (whether they be gases or liquids) in the heat exchanger.

This exchanger is used for cooling product with water which enters the shell side at 1 and leaves at 2. The product to be cooled flows through the small tubes. The exchanger has been taken out of service to clean the tubes to improve the efficiency of the cooling process.

Another example of a heat exchanger is the condenser at the top of a distillation column. Here heat from the vapour emerging from the top of the column is removed by, for example, cooling water. The vapour cools down and condenses and the temperature of the water increases.

Development of Chemical Reactor

In the future many chemicals may be produced in reactors about the size of a large desktop computer, known as microreactors. The reduced size will lead to a reduction in capital costs and a reduction in the amount of chemicals in use at any one time, resulting in an inherently safer process.

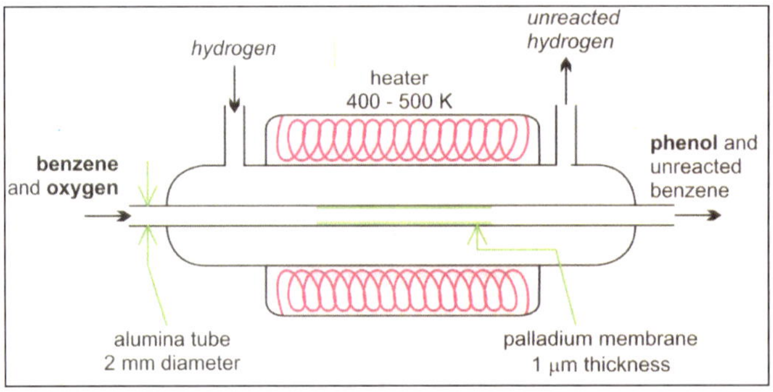

A microreactor being used to produce phenol from benzene.

The temperature can be kept constant more readily (as there is a much larger surface for a given volume). This allows for a very efficient heat transfer from the reaction to the surroundings even for very exothermic reactions such as the nitration of an aromatic hydrocarbon which can potentially be explosive.

There is considerable amount of research being carried out in developing microreactors. One example is the possibility of the direct conversion of benzene to phenol. A mixture of benzene and oxygen is fed through an alumina tube, packed with palladium at ca 350-400 K and hydrogen gas is passed over it.

Hydrogen permeates through the alumina tube, and is converted to atomic hydrogen by the palladium catalyst. The hydrogen atoms react with oxygen, releasing reactive oxygen species, such as hydroxyl radicals, which in turn react with the benzene to form phenol.

Another development is known as oscillatory flow mixing. Chemical engineers are designing reactors where the fluids to be reacted are oscillated inside a reactor with baffles at frequencies between 0.5 and 15 Hz with amplitudes in the range 1 to 100 mm. This allows for very effective mixing of the reactants and also for heat to be transferred to the surroundings. This gives similar conditions to those in plug flow which are otherwise difficult to achieve with small quantities of materials.

Chemical Plant

A chemical plant is an industrial process plant that manufactures (or otherwise processes) chemicals, usually on a large scale. The general objective of a chemical plant is to create new material wealth via the chemical or biological transformation and or separation of materials. Chemical plants use specialized equipment, units, and technology in the manufacturing process. Other kinds of plants, such as polymer, pharmaceutical, food, and some beverage production facilities, power plants, oil refineries or other refineries, natural gas processing and biochemical plants, water and wastewater treatment, and pollution control equipment use many technologies that have similarities to chemical plant technology such as fluid systems and chemical reactor systems. Some would consider an oil refinery or a pharmaceutical or polymer manufacturer to be effectively a chemical plant.

Petrochemical plants (plants using chemicals from petroleum as a raw material or *feedstock*) are usually located adjacent to an oil refinery to minimize transportation costs for the feedstocks produced by the refinery. Speciality chemical and fine chemical plants are usually much smaller and not as sensitive to location. Tools have been developed for converting a base project cost from one geographic location to another.

Chemical Processes

Chemical plants use chemical processes, which are detailed industrial-scale methods, to transform feedstock chemicals into products. The same chemical process can be used at more than one chemical plant, with possibly differently scaled capacities at each plant. Also, a chemical plant at a site may be constructed to utilize more than one chemical process, for instance to produce multiple products.

A chemical plant commonly has usually large vessels or sections called units or lines that are interconnected by piping or other material-moving equipment which can carry streams of material. Such material streams can include fluids (gas or liquid carried in piping) or sometimes solids or mixtures such as slurries. An overall chemical process is commonly made up of steps called unit operations which occur in the individual units. A raw material going into a chemical process or plant as input to be converted into a product is commonly called a feedstock, or simply feed. In addition to feedstocks for the plant as a whole, an input stream of material to be processed in a particular unit can similarly be considered feed for that unit. Output streams from the plant as a whole are final products and sometimes output streams from individual units may be considered intermediate products for their units. However, final products from one plant may be intermediate chemicals used as feedstock in another plant for further processing. For example, some products from an oil refinery may used as feedstock in petrochemical plants, which may in turn produce feedstocks for pharmaceutical plants.

Either the feedstock(s), the product(s), or both may be individual compounds or mixtures. It is often not worthwhile separating the components in these mixtures completely; specific levels of purity depend on product requirements and process economics.

Operations

Chemical processes may be run in continuous or batch operation.

Batch Operation

In batch operation, production occurs in time-sequential steps in discrete batches. A batch of feedstock(s) is fed (or charged) into a process or unit, then the chemical process takes place, then the product(s) and any other outputs are removed. Such batch production may be repeated over again and again with new batches of feedstock. Batch operation is commonly used in smaller scale plants such as pharmaceutical or specialty chemicals production, for purposes of improved traceability as well as flexibility. Continuous plants are usually used to manufacture commodity or petrochemicals while batch plants are more common in speciality and fine chemical production as well as pharmaceutical active ingredient (API) manufacture.

Continuous Operation

In continuous operation, all steps are ongoing continuously in time. During usual continuous operation, the feeding and product removal are ongoing streams of moving material, which together with the process itself, all take place simultaneously and continuously. Chemical plants or units in continuous operation are usually in a steady state or approximate steady state. Steady state means that quantities related to the process do not change as time passes during operation. Such constant quantities include stream flow rates, heating or cooling rates, temperatures, pressures, and chemical compositions at any given point (location). Continuous operation is more efficient in many large scale operations like petroleum refineries. It is possible for some units to operate continuously and others be in batch operation in a chemical plant. The amount of primary feedstock or product per unit of time which a plant or unit can process is referred to as the capacity of that plant or unit. For examples: the capacity of an oil refinery may be given in terms of barrels of crude oil refined per day; alternatively chemical plant capacity may be given in tons of product produced per day. In actual daily operation, a plant (or unit) will operate at a percentage of its full capacity. Engineers typically assume 90% operating time for plants which work primarily with fluids, and 80% uptime for plants which primarily work with solids.

Units and Fluid Systems

Specific unit operations are conducted in specific kinds of units. Although some units may operate at ambient temperature or pressure, many units operate at higher or lower temperatures or pressures. Vessels in chemical plants are often cylindrical with rounded ends, a shape which can be suited to hold either high pressure or vacuum. Chemical reactions can convert certain kinds of compounds into other compounds in chemical reactors. Chemical reactors may be packed beds and may have solid heterogeneous catalysts which stay in the reactors as fluids move through, or may simply be stirred vessels in which reactions occur. Since the surface of solid heterogeneous catalysts may sometimes become "poisoned" from deposits such as coke, regeneration of catalysts may be necessary. Fluidized beds may also be used in some cases to ensure good mixing. There can also be units (or subunits) for mixing (including dissolving), separation, heating, cooling, or some combination of these. For example, chemical reactors often have stirring for mixing and heating or cooling to maintain temperature. When designing plants on a large scale, heat produced or absorbed by chemical reactions must be considered. Some plants may have units with organism cultures for biochemical processes such as fermentation or enzyme production.

Separation processes include filtration, settling (sedimentation), extraction or leaching, distillation, recrystallization or precipitation (followed by filtration or settling), reverse osmosis, drying, and adsorption. Heat exchangers are often used for heating or cooling, including boiling or condensation, often in conjunction with other units such as distillation towers. There may also be storage tanks for storing feedstock, intermediate or final products, or waste. Storage tanks commonly have level indicators to show how full they are. There may be structures holding or supporting sometimes massive units and their associated equipment. There are often stairs, ladders, or other steps for personnel to reach points in the units for sampling, inspection, or maintenance. An area of a plant or facility with numerous storage tanks is sometimes called a *tank farm*, especially at an oil depot.

Distillation unit.

Fluid systems for carrying liquids and gases include piping and tubing of various diameter sizes, various types of valves for controlling or stopping flow, pumps for moving or pressurizing liquid, and compressors for pressurizing or moving gases. Vessels, piping, tubing, and sometimes other equipment at high or very low temperature are commonly covered with insulation for personnel safety and to maintain temperature inside. Fluid systems and units commonly have instrumentation such as temperature and pressure sensors and flow measuring devices at select locations in a plant. Online analyzers for chemical or physical property analysis have become more common. Solvents can sometimes be used to dissolve reactants or materials such as solids for extraction or leaching, to provide a suitable medium for certain chemical reactions to run, or so they can otherwise be treated as fluids.

Chemical Plant Design

Today, the fundamental aspects of designing chemical plants are done by chemical engineers. Historically, this was not always the case and many chemical plants were constructed in a haphazard way before the discipline of chemical engineering became established. Chemical engineering was first established as a profession in the United Kingdom when the first chemical engineering course was given at the University of Manchester in 1887 by George E. Davis in the form of twelve lectures covering various aspects of industrial chemical practice. As a consequence George E. Davis is regarded as the World's first Chemical Engineer. Today Chemical Engineering is a profession and those Professional Chemical Engineers with experience can gain "Chartered" engineer status through the Institution of Chemical Engineers.

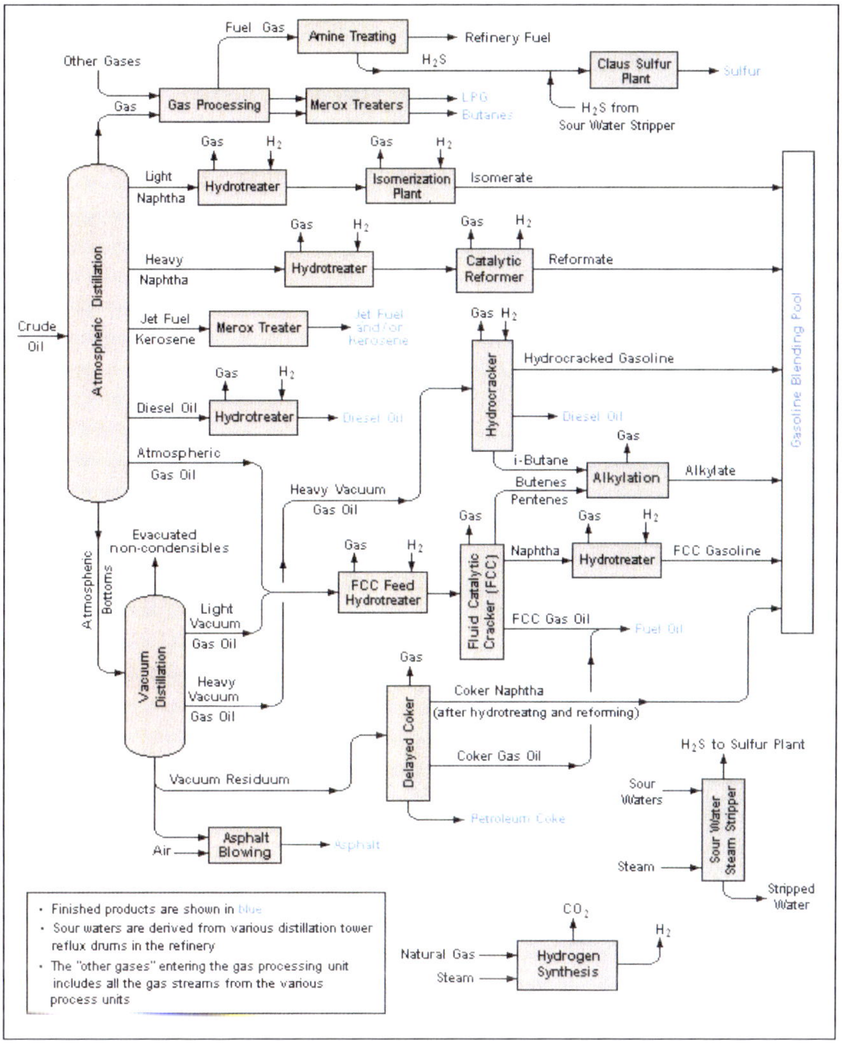

Flow diagram for a typical oil refinery.

In plant design, typically less than 1 per cent of ideas for new designs ever become commercialized. During this solution process, typically, cost studies are used as an initial screening to eliminate unprofitable designs. If a process appears profitable, then other factors are considered, such as safety, environmental constraints, controllability, etc. The general goal in plant design, is to construct or synthesize "optimum designs" in the neighborhood of the desired constraints.

Many times chemists research chemical reactions or other chemical principles in a laboratory, commonly on a small scale in a "batch-type" experiment. Chemistry information obtained is then used by chemical engineers, along with expertise of their own, to convert to a chemical process and scale up the batch size or capacity. Commonly, a small chemical plant called a pilot plant is built to provide design and operating information before construction of a large plant. From data and operating experience obtained from the pilot plant, a scaled-up plant can be designed for higher or full capacity. After the fundamental aspects of a plant design are determined, mechanical or electrical engineers may become involved with mechanical or electrical details, respectively. Structural engineers may become involved in the plant design to ensure the structures can support the weight of the units, piping, and other equipment.

The units, streams, and fluid systems of chemical plants or processes can be represented by block flow diagrams which are very simplified diagrams, or process flow diagrams which are somewhat more detailed. The streams and other piping are shown as lines with arrow heads showing usual direction of material flow. In block diagrams, units are often simply shown as blocks. Process flow diagrams may use more detailed symbols and show pumps, compressors, and major valves. Likely values or ranges of material flow rates for the various streams are determined based on desired plant capacity using material balance calculations. Energy balances are also done based on heats of reaction, heat capacities, expected temperatures and pressures at various points to calculate amounts of heating and cooling needed in various places and to size heat exchangers. Chemical plant design can be shown in fuller detail in a piping and instrumentation diagram (P&ID) which shows all piping, tubing, valves, and instrumentation, typically with special symbols. Showing a full plant is often complicated in a P&ID, so often only individual units or specific fluid systems are shown in a single P&ID.

In the plant design, the units are sized for the maximum capacity each may have to handle. Similarly, sizes for pipes, pumps, compressors, and associated equipment are chosen for the flow capacity they have to handle. Utility systems such as electric power and water supply should also be included in the plant design. Additional piping lines for non-routine or alternate operating procedures, such as plant or unit startups and shutdowns, may have to be included. Fluid systems design commonly includes isolation valves around various units or parts of a plant so that a section of a plant could be isolated in case of a problem such as a leak in a unit. If pneumatically or hydraulically actuated valves are used, a system of pressurizing lines to the actuators is needed. Any points where process samples may have to be taken should have sampling lines, valves, and access to them included in the detailed design. If necessary, provisions should be made for reducing high pressure or temperature of a sampling stream, such including a pressure reducing valve or sample cooler.

Units and fluid systems in the plant including all vessels, piping, tubing, valves, pumps, compressors, and other equipment must be rated or designed to be able to withstand the entire range of pressures, temperatures, and other conditions which they could possibly encounter, including any appropriate safety factors. All such units and equipment should also be checked for materials compatibility to ensure they can withstand long-term exposure to the chemicals they will come in contact with. Any closed system in a plant which has a means of pressurizing possibly beyond the rating of its equipment, such as heating, exothermic reactions, or certain pumps or compressors, should have an appropriately sized pressure relief valve included to prevent overpressurization for safety. Frequently all of these parameters (temperatures, pressures, flow, etc.) are exhaustively analyzed in combination through a *Hazop* or *fault tree analysis*, to ensure that the plant has no known risk of serious hazard.

Within any constraints the plant is subject to, design parameters are optimized for good economic performance while ensuring safety and welfare of personnel and the surrounding community. For flexibility, a plant may be designed to operate in a range around some optimal design parameters in case feedstock or economic conditions change and re-optimization is desirable. In more modern times, computer simulations or other computer calculations have been used to help in chemical plant design or optimization.

Plant Operation

Process Control

In process control, information gathered automatically from various sensors or other devices in the plant is used to control various equipment for running the plant, thereby controlling operation of the plant. Instruments receiving such information signals and sending out control signals to perform this function automatically are process *controllers*. Previously, pneumatic controls were sometimes used. Electrical controls are now common. A plant often has a control room with displays of parameters such as key temperatures, pressures, fluid flow rates and levels, operating positions of key valves, pumps and other equipment, etc. In addition, operators in the control room can control various aspects of the plant operation, often including overriding automatic control. Process control with a computer represents more modern technology. Based on possible changing feedstock composition, changing products requirements or economics, or other changes in constraints, operating conditions may be re-optimized to maximize profit.

Workers

As in any industrial setting, there are a variety of workers working throughout a chemical plant facility, often organized into departments, sections, or other work groups. Such workers typically include engineers, plant operators, and maintenance technicians. Other personnel at the site could include chemists, management/administration and office workers. Types of engineers involved in operations or maintenance may include chemical process engineers, mechanical engineers for maintaining mechanical equipment, and electrical/computer engineers for electrical or computer equipment.

Transport

Large quantities of fluid feedstock or product may enter or leave a plant by pipeline, railroad tank car, or tanker truck. For example, petroleum commonly comes to a refinery by pipeline. Pipelines can also carry petrochemical feedstock from a refinery to a nearby petrochemical plant. Natural gas is a product which comes all the way from a natural gas processing plant to final consumers by pipeline or tubing. Large quantities of liquid feedstock are typically pumped into process units. Smaller quantities of feedstock or product may be shipped to or from a plant in drums. Use of drums about 55 gallons in capacity is common for packaging industrial quantities of chemicals. Smaller batches of feedstock may be added from drums or other containers to process units by workers.

Maintenance

In addition to feeding and operating the plant, and packaging or preparing the product for shipping, plant workers are needed for taking samples for routine and troubleshooting analysis and for performing routine and non-routine maintenance. Routine maintenance can include periodic inspections and replacement of worn catalyst, analyzer reagents, various sensors, or mechanical parts. Non-routine maintenance can include investigating problems and then fixing them, such as leaks, failure to meet feed or product specifications, mechanical failures of valves, pumps, compressors, sensors, etc.

Statutory and Regulatory Compliance

When working with chemicals, safety is a concern in order to avoid problems such as chemical accidents. In the United States, the law requires that employers provide workers working with chemicals with access to a Material Safety Data Sheet (MSDS) for every kind of chemical they work with. An MSDS for a certain chemical is prepared and provided by the supplier to whoever buys the chemical. Other laws covering chemical safety, hazardous waste, and pollution must be observed, including statutes such as the Resource Conservation and Recovery Act (RCRA) and the Toxic Substances Control Act (TSCA), and regulations such as the Chemical Facility Anti-Terrorism Standards in the United States. Hazmat (hazardous materials) teams are trained to deal with chemical leaks or spills. Process Hazard Analysis (PHA) is used to assess potential hazards in chemical plants. In 1998, the U. S. Chemical Safety and Hazard Investigation Board has become operational.

Plant Facilities

The actual production or process part of a plant may be indoors, outdoors, or a combination of the two. It may be a traditional stick-built plant or a modular skid. Large modular skids are especially impressive feats of engineering. A modular skid is built including all of the modular equipment needed to do the same job a traditional stick-build plant may perform. However, the modular skid is built within a structural steel frame, allowing it to be shipped to the onsite location without needing to be rebuilt onsite. A modular skid build results in a higher functioning end product, as less hands are required in the onsite setup of the modular skid process unit, resulting in minimized risk for mishaps. The actual production section of a facility usually has the appearance of a rather industrial environment. Hard hats and work shoes are commonly worn. Floors and stairs are often made of metal grating, and there is practically no decoration. There may also be pollution control or waste treatment facilities or equipment. Sometimes existing plants may be expanded or modified based on changing economics, feedstock, or product needs. As in other production facilities, there may be shipping and receiving, and storage facilities. In addition, there are usually certain other facilities, typically indoors, to support production at the site.

Although some simple sample analysis may be able to be done by operations technicians in the plant area, a chemical plant typically has a laboratory where chemists analyze samples taken from the plant. Such analysis can include chemical analysis or determination of physical properties. Sample analysis can include routine quality control on feedstock coming into the plant, intermediate and final products to ensure quality specifications are met. Non-routine samples may be taken and analyzed for investigating plant process problems also. A larger chemical company often has a research laboratory for developing and testing products and processes where there may be pilot plants, but such a laboratory may be located at a site separate from the production plants.

A plant may also have a workshop or maintenance facility for repairs or keeping maintenance equipment. There is also typically some office space for engineers, management or administration, and perhaps for receiving visitors. The decorum there is commonly more typical of an office environment.

Clustering of Commodity Chemical Plants

Chemical Plants used particularly for commodity chemical and petrochemical manufacture, are located in relatively few manufacturing locations around the world largely due to infrastructural needs. This is less important for speciality or fine chemical batch plants. Not all commodity/petrochemicals are produced in any one location but groups of related materials often are, to induce industrial symbiosis as well as material, energy and utility efficiency and other economies of scale. These manufacturing locations often have business clusters of units called chemical plants that share utilities and large scale infrastructure such as power stations, port facilities, road and rail terminals. In the United Kingdom for example there are four main locations for commodity chemical manufacture: near the River Mersey in Northwest England, on the Humber on the East coast of Yorkshire, in Grangemouth near the Firth of Forth in Scotland and on Teesside as part of the Northeast of England Process Industry Cluster (NEPIC). Approximately 50% of the UK's petrochemicals, which are also commodity chemicals, are produced by the industry cluster companies on Teesside at the mouth of the River Tees on three large chemical parks at Wilton, Billingham and Seal Sands.

Corrosion and use of New Materials

Corrosion in chemical process plants is a major issue that consumes billions of dollars yearly. Electrochemical corrosion of metals is pronounced in chemical process plants due to the presence of acid fumes and other electrolytic interactions. Recently, FRP (Fibre-reinforced plastic) is used as a material of construction. The British standard specification BS4994 is widely used for design and construction of the vessels, tanks, etc.

Transport Phenomena

The study of transport phenomena concerns the exchange of mass, energy, charge, momentum and angular momentum between observed and studied systems. While it draws from fields as diverse as continuum mechanics and thermodynamics, it places a heavy emphasis on the commonalities between the topics covered. Mass, momentum, and heat transport all share a very similar mathematical framework, and the parallels between them are exploited in the study of transport phenomena to draw deep mathematical connections that often provide very useful tools in the analysis of one field that are directly derived from the others.

The fundamental analysis in all three subfields of mass, heat, and momentum transfer are often grounded in the simple principle that the sum total of the quantities being studied must be conserved by the system and its environment. Thus, the different phenomena that lead to transport are each considered individually with the knowledge that the sum of their contributions must equal zero. This principle is useful for calculating many relevant quantities. For example, in fluid mechanics, a common use of transport analysis is to determine the velocity profile of a fluid flowing through a rigid volume.

Transport phenomena are ubiquitous throughout the engineering disciplines. Some of the most common examples of transport analysis in engineering are seen in the fields of process, chemical, biological, and mechanical engineering, but the subject is a fundamental component of the curriculum in all disciplines involved in any way with fluid mechanics, heat transfer, and mass transfer. It is now considered to be a part of the engineering discipline as much as thermodynamics, mechanics, and electromagnetism.

Transport phenomena encompass all agents of physical change in the universe. Moreover, they are considered to be fundamental building blocks which developed the universe, and which is responsible for the success of all life on earth. However, the scope here is limited to the relationship of transport phenomena to artificial engineered systems.

In physics, transport phenomena are all irreversible processes of statistical nature stemming from the random continuous motion of molecules, mostly observed in fluids.

Every aspect of transport phenomena is grounded in two primary concepts : the conservation laws, and the constitutive equations. The conservation laws, which in the context of transport phenomena are formulated as continuity equations, describe how the quantity being studied must be conserved. The constitutive equations describe how the quantity in question responds to various stimuli via transport. Prominent examples include Fourier's Law of Heat Conduction and the Navier-Stokes equations, which describe, respectively, the response of heat flux to temperature gradients and the relationship between fluid flux and the forces applied to the fluid. These equations also demonstrate the deep connection between transport phenomena and thermodynamics, a connection that explains why transport phenomena are irreversible. Almost all of these physical phenomena ultimately involve systems seeking their lowest energy state in keeping with the principle of minimum energy. As they approach this state, they tend to achieve true thermodynamic equilibrium, at which point there are no longer any driving forces in the system and transport ceases. The various aspects of such equilibrium are directly connected to a specific transport: heat transfer is the system's attempt to achieve thermal equilibrium with its environment, just as mass and momentum transport move the system towards chemical and mechanical equilibrium.

Examples of transport processes include heat conduction (energy transfer), fluid flow (momentum transfer), molecular diffusion (mass transfer), radiation and electric charge transfer in semiconductors.

Transport phenomena have wide application. For example, in solid state physics, the motion and interaction of electrons, holes and phonons are studied under "transport phenomena". Another example is in biomedical engineering, where some transport phenomena of interest are thermoregulation, perfusion, and microfluidics. In chemical engineering, transport phenomena are studied in reactor design, analysis of molecular or diffusive transport mechanisms, and metallurgy.

The transport of mass, energy, and momentum can be affected by the presence of external sources:

- An odor dissipates more slowly (and may intensify) when the source of the odor remains present.
- The rate of cooling of a solid that is conducting heat depends on whether a heat source is applied.
- The gravitational force acting on a rain drop counteracts the resistance or drag imparted by the surrounding air.

Commonalities among Phenomena

An important principle in the study of transport phenomena is analogy between phenomena.

Diffusion

There are some notable similarities in equations for momentum, energy, and mass transfer which can all be transported by diffusion, as illustrated by the following examples:

- Mass: The spreading and dissipation of odors in air is an example of mass diffusion.
- Energy: The conduction of heat in a solid material is an example of heat diffusion.
- Momentum: The drag experienced by a rain drop as it falls in the atmosphere is an example of momentum diffusion (the rain drop loses momentum to the surrounding air through viscous stresses and decelerates).

The molecular transfer equations of Newton's law for fluid momentum, Fourier's law for heat, and Fick's law for mass are very similar. One can convert from one transfer coefficient to another in order to compare all three different transport phenomena.

Comparison of diffusion phenomena		
Transported quantity	Physical phenomenon	Equation
Momentum	Viscosity (Newtonian fluid)	$\tau = -\upsilon \frac{\partial \rho \upsilon}{\partial x}$
Energy	Heat conduction (Fourier's law)	$\frac{q}{A} = -k \frac{dT}{dx}$
Mass	Molecular diffusion (Fick's law)	$J = -D \frac{\partial C}{\partial x}$

A great deal of effort has been devoted in the literature to developing analogies among these three transport processes for turbulent transfer so as to allow prediction of one from any of the others. The Reynolds analogy assumes that the turbulent diffusivities are all equal and that the molecular diffusivities of momentum (μ/ρ) and mass (D_{AB}) are negligible compared to the turbulent diffusivities. When liquids are present and/or drag is present, the analogy is not valid. Other analogies, such as von Karman's and Prandtl's, usually result in poor relations.

The most successful and most widely used analogy is the Chilton and Colburn J-factor analogy. This analogy is based on experimental data for gases and liquids in both the laminar and turbulent regimes. Although it is based on experimental data, it can be shown to satisfy the exact solution derived from laminar flow over a flat plate. All of this information is used to predict transfer of mass.

Onsager Reciprocal Relations

In fluid systems described in terms of temperature, matter density, and pressure, it is known that temperature differences lead to heat flows from the warmer to the colder parts of the system; similarly, pressure differences will lead to matter flow from high-pressure to low-pressure regions (a "reciprocal relation"). What is remarkable is the observation that, when both pressure and temperature vary, temperature differences at constant pressure can cause matter flow (as in convection) and pressure differences at constant temperature can cause heat flow. Perhaps surprisingly, the heat flow per unit of pressure difference and the density (matter) flow per unit of temperature difference are equal.

This equality was shown to be necessary by Lars Onsager using statistical mechanics as a consequence of the time reversibility of microscopic dynamics. The theory developed by Onsager is much more general than this example and capable of treating more than two thermodynamic forces at once.

Momentum Transfer

In momentum transfer, the fluid is treated as a continuous distribution of matter. The study of momentum transfer, or fluid mechanics can be divided into two branches: fluid statics (fluids at rest), and fluid dynamics (fluids in motion). When a fluid is flowing in the x-direction parallel to a solid surface, the fluid has x-directed momentum, and its concentration is $\upsilon_x\rho$. By random diffusion of molecules there is an exchange of molecules in the z-direction. Hence the x-directed momentum has been transferred in the z-direction from the faster- to the slower-moving layer. The equation for momentum transport is Newton's Law of Viscosity written as follows:

$$\tau_{zx} = -\upsilon \frac{\partial \rho \upsilon_x}{\partial z}$$

where τ_{zx} is the flux of x-directed momentum in the z-direction, v is μ/ρ, the momentum diffusivity, z is the distance of transport or diffusion, ρ is the density, and μ is the dynamic viscosity. Newton's Law is the simplest relationship between the flux of momentum and the velocity gradient.

Heat and Mass Transfer

When a system contains two or more components whose concentration vary from point to point, there is a natural tendency for mass to be transferred, minimizing any concentration difference within the system. Mass Transfer in a system is governed by Fick's First Law: 'Diffusion flux from higher concentration to lower concentration is proportional to the gradient of the concentration of the substance and the diffusivity of the substance in the medium.' Mass transfer can take place due to different driving forces. Some of them are:

- Mass can be transferred by the action of a pressure gradient (pressure diffusion).
- Forced diffusion occurs because of the action of some external force.
- Diffusion can be caused by temperature gradients (thermal diffusion).
- Diffusion can be caused by differences in chemical potential.

This can be compared to Fick's Law of Diffusion, for a species A in a binary mixture consisting of A and B:

$$J_{Ay} = -D_{AB}\frac{\partial Ca}{\partial y}$$

where, D is the diffusivity constant.

Energy Transfer

All processes in engineering involve the transfer of energy. Some examples are the heating and cooling of process streams, phase changes, distillations, etc. The basic principle is the first law of thermodynamics which is expressed as follows for a static system:

$$q = -k\frac{dT}{dx}$$

The net flux of energy through a system equals the conductivity times the rate of change of temperature with respect to position.

For other systems that involve either turbulent flow, complex geometries or difficult boundary conditions another equation would be easier to use:

$$Q = h \cdot A \cdot \Delta T$$

where A is the surface area, : ΔT is the temperature driving force, Q is the heat flow per unit time, and h is the heat transfer coefficient.

Within heat transfer, two types of convection can occur:

Forced convection can occur in both laminar and turbulent flow. In the situation of laminar flow in circular tubes, several dimensionless numbers are used such as Nusselt number, Reynolds number, and Prandtl. The commonly used equation is:

$$Nu_a = \frac{h_a D}{k}$$

Natural or free convection is a function of Grashof and Prandtl numbers. The complexities of free convection heat transfer make it necessary to mainly use empirical relations from experimental data. Heat transfer is analyzed in packed beds, reactors and heat exchangers.

Chemical Process

In a scientific sense, a chemical process is a method or means of somehow changing one or more chemicals or chemical compounds. Such a chemical process can occur by itself or be caused by an outside force, and involves a chemical reaction of some sort. In an "engineering" sense, a chemical process is a method intended to be used in manufacturing or on an industrial scale to change the composition of chemical(s) or material(s), usually using technology similar or related to that used in chemical plants or the chemical industry.

Neither of these definitions are exact in the sense that one can always tell definitively what is a chemical process and what is not; they are practical definitions. There is also significant overlap in these two definition variations. Because of the inexactness of the definition, chemists and other scientists use the term "chemical process" only in a general sense or in the engineering sense. However, in the "process (engineering)" sense, the term "chemical process" is used extensively.

Although this type of chemical process may sometimes involve only one step, often multiple steps, referred to as unit operations, are involved. In a plant, each of the unit operations commonly occur in individual vessels or sections of the plant called units. Often, one or more chemical reactions are involved, but other ways of changing chemical (or material) composition may be used, such as mixing or separation processes. The process steps may be sequential in time or sequential in space along a stream of flowing or moving material. For a given amount of a feed (input) material or product (output) material, an expected amount of material can be determined at key steps in the process from empirical data and material balance calculations. These amounts can be scaled up or down to suit the desired capacity or operation of a particular chemical plant built for such a process. More than one chemical plant may use the same chemical process, each plant perhaps at differently scaled capacities. Chemical processes like distillation and crystallization go back to alchemy in Alexandria, Egypt.

Such chemical processes can be illustrated generally as block flow diagrams or in more detail as process flow diagrams. Block flow diagrams show the units as blocks and the streams flowing between them as connecting lines with arrowheads to show direction of flow.

In addition to chemical plants for producing chemicals, chemical processes with similar technology and equipment are also used in oil refining and other refineries, natural gas processing, polymer and pharmaceutical manufacturing, food processing, and water and wastewater treatment.

Unit Processing in Chemical Process

Unit processing is the basic processing in chemical engineering. Together with unit operations it forms the main principle of the varied chemical industries. Each genre of unit processing follows the same chemical law much as each genre of unit operations follows the same physical law.

Chemical engineering unit processing consists of the following important processes:

- Oxidation.
- Reduction.
- Hydrogenation.
- Dehydrogenation.
- Hydrolysis.
- Hydration.
- Dehydration.
- Halogenation.
- Nitrification.
- Sulfonation.
- Ammoniation.
- Alkaline fusion.
- Alkylation.
- Dealkylation.
- Esterification.
- Polymerization
- Polycondensation.
- Catalysis.

Unit Operation

In chemical engineering and related fields, a unit operation is a basic step in a process. Unit operations involve a physical change or chemical transformation such as separation, crystallization, evaporation, filtration, polymerization, isomerization, and other reactions. For example, in milk processing, homogenization, pasteurization, and packaging are each unit operations which are connected to create the overall process. A process may require many unit operations to obtain the desired product from the starting materials, or feedstocks.

Chemical engineering unit operations consist of five classes:

- Fluid flow processes, including fluids transportation, filtration, and solids fluidization.
- Heat transfer processes, including evaporation and heat exchange.

- Mass transfer processes, including gas absorption, distillation, extraction, adsorption, and drying.
- Thermodynamic processes, including gas liquefaction, and refrigeration.
- Mechanical processes, including solids transportation, crushing and pulverization, and screening and sieving.

Chemical engineering unit operations also fall in the following categories which involve elements from more than one class:

- Combination (mixing).
- Separation (distillation, crystallization).
- Reaction (chemical reaction).

Furthermore, there are some unit operations which combine even these categories, such as reactive distillation and stirred tank reactors. A "pure" unit operation is a physical transport process, while a mixed chemical/physical process requires modeling both the physical transport, such as diffusion, *and* the chemical reaction. This is usually necessary for designing catalytic reactions, and is considered a separate discipline, termed chemical reaction engineering.

Chemical engineering unit operations and chemical engineering unit processing form the main principles of all kinds of chemical industries and are the foundation of designs of chemical plants, factories, and equipment used.

In general, unit operations are designed by writing down the balances for the transported quantity for each elementary component (which may be infinitesimal) in the form of equations, and solving the equations for the design parameters, then selecting an optimal solution out of the several possible and then designing the physical equipment. For instance, distillation in a plate column is analyzed by writing down the mass balances for each plate, wherein the known vapor-liquid equilibrium and efficiency, drip out and drip in comprise the total mass flows, with a sub-flow for each component. Combining a stack of these gives the system of equations for the whole column. There is a range of solutions, because a higher reflux ratio enables fewer plates, and vice versa. The engineer must then find the optimal solution with respect to acceptable volume holdup, column height and cost of construction.

Process Design

In chemical engineering, process design is the choice and sequencing of units for desired physical and/or chemical transformation of materials. Process design is central to chemical engineering, and it can be considered to be the summit of that field, bringing together all of the field's components.

Process design can be the design of new facilities or it can be the modification or expansion of existing facilities. The design starts at a conceptual level and ultimately ends in the form of fabrication and construction plans.

Process design is distinct from equipment design, which is closer in spirit to the design of unit operations. Processes often include many unit operations.

Documentation

Process design documents serve to define the design and they ensure that the design components fit together. They are useful in communicating ideas and plans to other engineers involved with the design, to external regulatory agencies, to equipment vendors and to construction contractors.

In order of increasing detail, process design documents include:

- Block flow diagrams (BFD): Very simple diagrams composed of rectangles and lines indicating major material or energy flows.
- Process flow diagrams (PFD): Typically more complex diagrams of major unit operations as well as flow lines. They usually include a material balance, and sometimes an energy balance, showing typical or design flowrates, stream compositions, and stream and equipment pressures and temperatures.
- Piping and instrumentation diagrams (P&ID): Diagrams showing each and every pipeline with piping class (carbon steel or stainless steel) and pipe size (diameter). They also show valving along with instrument locations and process control schemes.
- Specifications: Written design requirements of all major equipment items.

Process designers typically write operating manuals on how to start-up, operate and shut-down the process. They often also develop accident plans and projections of process operation on the environment.

Documents are maintained after construction of the process facility for the operating personnel to refer to. The documents also are useful when modifications to the facility are planned. A primary method of developing the process documents is process flowsheeting.

Design Considerations

There are several considerations that need to be made when designing any chemical process unit. Design conceptualization and considerations can begin once product purities, yields, and throughput rates are all defined.

Objectives that a design may strive to include:

- Throughput rate.
- Process yield.
- Product purity.

Constraints include:

- Capital cost: the amount of budget or investment to construct end to end process.
- Available space: the area of the land to build the plant.
- Safety concerns: consideration towards risk analysis on industrial accidents or hazardous chemicals.
- Environmental impact and projected effluents and emissions.
- Waste production/recycling: manage waste produced as side product of the process for not to harm the surroundings.
- Operating and maintenance costs: represent the variable cost of the operational of the plant.

Other factors that designers may include are:

- Reliability.
- Redundancy.
- Flexibility.
- Anticipated variability in feed stock and allowable variability in product.

Sources of Design Information

Designers usually do not start from scratch, especially for complex projects. Often the engineers have pilot plant data available or data from full-scale operating facilities. Other sources of information include proprietary design criteria provided by process licensors, published scientific data, laboratory experiments, and suppliers of feedstocks and utilities.

Design Process

Design starts with process synthesis - the choice of technology and combinations of industrial units to achieve goals. More detailed design proceeds as other engineers and stakeholders sign off on each stage: conceptual to detailed design.

Simulation software is often used by design engineers. Simulations can identify weaknesses in designs and allow engineers to choose better alternatives. However, engineers still rely on heuristics, intuition, and experience when designing a process. Human creativity is an element in complex designs.

Separation Process

A separation process is a method that converts a mixture or solution of chemical substances into two or more distinct product mixtures. At least one of results of the separation is enriched in one or more of the source mixture's constituents. In some cases, a separation may fully divide the mixture into pure constituents. Separations exploit differences in chemical properties or physical properties (such as size, shape, mass, density, or chemical affinity) between the constituents of a mixture.

Processes are often classified according to the particular differences they use to achieve separation. If no single difference can be used to accomplish a desired separation, multiple operations can often be combined to achieve the desired end.

With a few exceptions, elements or compounds exist in nature in an impure state. Often these raw materials must go through a separation before they can be put to productive use, making separation techniques essential for the modern industrial economy.

The purpose of a separation may be *analytical*, can be used as a lie components in the original mixture without any attempt to save the fractions, or may be *preparative*, i.e. to "prepare" fractions or samples of the components that can be saved. The separation can be done on a small scale, effectively a laboratory scale for analytical or preparative purposes, or on a large scale, effectively an industrial scale for preparative purposes, or on some intermediate scale.

Complete and Incomplete Separation

In some cases, separations require total purification, as in the electrolysis refining of bauxite ore for aluminum metal, but a good example of an incomplete separation technique is oil refining. Crude oil occurs naturally as a mixture of various hydrocarbons and impurities. The refining process splits this mixture into other, more valuable mixtures such as natural gas, gasoline and chemical feedstocks, none of which are pure substances, but each of which must be separated from the raw crude. In both of these cases, a series of separations is necessary to obtain the desired end products. In the case of oil refining, crude is subjected to a long series of individual distillation steps, each of which produces a different product or intermediate.

Separating Liquids

Separators are used to divide liquids. Vertically supported centrifuges are built with flying bearings. A separator is a continuous sedimentation centrifuge. Both exit streams are continuously discharged, using a pump (under pressure) or pressure free. The solid material can be discharged discontinuously (chamber drum, solid walled disc drum), pseudo continuously (self-cleaning disc drum) or continuously (nozzle drum). The drum is the centerpiece of the separator, in which the separation process takes place. There are two types of drums: the chamber drum (known as chamber separators) and the disc drum (known as disc separators). The power transmission on the spindle and thereby on the drum can take place by using one of the three drive motors: helical gears, a belt drive or direct drive, via a special motor. The sealing of the separators is differentiated into four types: open, semi-closed, hydro-hermetic (sealing of the product space) or fully hermetic (absolute airtight).

Distillation

Distillation is a method of separating chemical substances based on differences in their volatilities. It usually forms part of a larger chemical process, and is thus referred to as a unit operation.

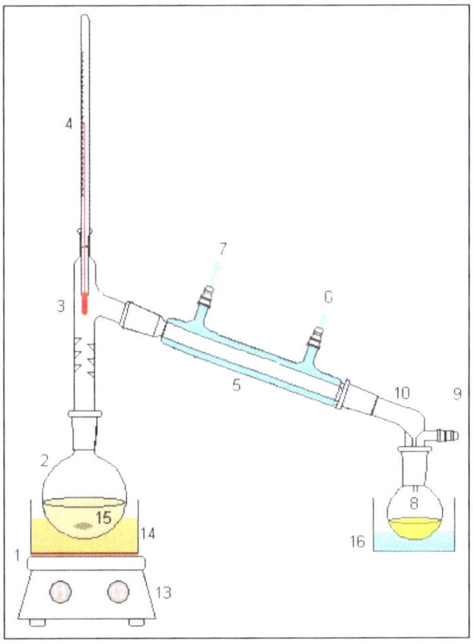

A simple distillation apparatus used in a laboratory.

In the figure, 1: Heat source 2: Still pot 3: Still head 4: Thermometer/Boiling point temperature 5: Condenser 6: Cooling water in 7: Cooling water out 8: Distillate/receiving flask 9: Vacuum/gas inlet 10: Still receiver 11: Heat control 12: Stirrer speed control 13: Stirrer/heat plate 14: Heating (Oil/sand) bath 15: Stirrer bar/anti-bumping granules 16: Cooling bath.

Distillation is used for a variety of purposes. Water is distilled to remove impurities—for example, to remove salt from seawater. Air is distilled to separate its components, such as oxygen, which is used for medical applications, and helium, used for party balloons. The distillation of crude oil produces fractions useful for specific purposes such as transport, power generation, and heating. The use of distillation on fermented solutions to produce distilled beverages with a higher alcohol content is perhaps the oldest form of distillation, known since ancient times.

Applications of Distillation

The application of distillation can roughly be divided in four groups: laboratory scale, industrial distillation, distillation of herbs for perfumery and medicinals (herbal distillate) and food processing. The latter two are distinct from the former two, in that in the distillation is not used as a true purification method, but more to transfer all volatiles from the source materials to the distillate.

The main difference between laboratory scale distillation and industrial distillation is that laboratory scale distillation is often performed batch-wise, whereas industrial distillation often occurs continuously. In batch distillation, the composition of the source material, the vapors of the distilling compounds and the distillate change during the distillation. In batch distillation, a still is charged (supplied) with a batch of feed mixture, which is then separated into its component fractions which are collected sequentially from most volatile to less volatile, with the bottoms (remaining least or non-volatile fraction) removed at the end. The still can then be recharged and the process repeated.

In continuous distillation, the source materials, vapors and distillate are kept at a constant composition by carefully replenishing the source material and removing fractions from both vapor and liquid in the system. This results in a better control of the separation process.

Idealized Distillation Model

It is a common misconception that in a solution, each component boils at its normal boiling point - the vapors of each component will collect separately and purely. This, however, does not occur even in an idealized system. Idealized models of distillation are essentially governed by Raoult's law and Dalton's law.

Raoult's law assumes that a component contributes to the total vapor pressure of the mixture in proportion to its percentage of the mixture and its vapor pressure when pure. If one component changes another component's vapor pressure, or if the volatility of a component is dependent on its percentage in the mixture, the law will fail.

Dalton's law states that the total vapor pressure is the sum of the vapor pressures of each individual component in the mixture. When a multi-component system is heated, the vapor pressure of each component will rise, thus causing the total vapor pressure to rise. When the total vapor pressure reaches the ambient pressure, boiling occurs and liquid turns to gas throughout the bulk of the solution. Note that a given mixture has one boiling point, when the components are mutually soluble.

The idealized model is accurate in the case of chemically similar liquids, such as benzene and toluene. In other cases, severe deviations from Raoult's law and Dalton's law are observed, most famously in the mixture of ethanol and water. These compounds, when heated together, form an azeotrope, in which the boiling temperature of the mixture is lower than the boiling temperature of each separate liquid. Virtually all liquids, when mixed and heated, will display azeotropic behavior. Although there are computational methods that can be used to estimate the behavior of a mixture of arbitrary components, the only way to obtain accurate vapor-liquid equilibrium data is by measurement.

It is not possible to completely purify a mixture of components by distillation, as this would require each component in the mixture to have a zero partial pressure. If ultra-pure products are the goal, then further chemical separation must be applied.

Batch Distillation

Heating an ideal mixture of two volatile substances A and B (with A having the higher volatility, or lower boiling point) in a batch distillation setup until the mixture is boiling results in a vapor above the liquid which contains a mixture of A and B. The ratio between A and B in the vapor will be different from the ratio in the liquid: the ratio in the liquid will be determined by how the original mixture was prepared, while the ratio in the vapor will be enriched in the more volatile compound, A (due to Raoult's Law). The vapor goes through the condenser and is removed from the system. This in turn means that the ratio of compounds in the remaining liquid is now different from the initial ratio (i.e. more enriched in B than the starting liquid).

The result is that the ratio in the liquid mixture is changing, becoming richer in component B. This causes the boiling point of the mixture to rise, which in turn results in a rise in the temperature in the vapor, which results in a changing ratio of A : B in the gas phase (as distillation continues, there is an increasing proportion of B in the gas phase). This results in a slowly changing ratio A : B in the distillate.

If the difference in vapor pressure between the two components A and B is large (generally expressed as the difference in boiling points), the mixture in the beginning of the distillation is highly enriched in component A, and when component A has distilled off, the boiling liquid is enriched in component B.

Continuous Distillation

In continuous distillation, the process is different from the above in that fractions are withdrawn from both the vapor and the liquid at such a speed that the combined ratio of the two fractions is exactly the same as the ratio in the starting mixture. In this way a stream of enriched component A and a stream of enriched component B is obtained. Moreover, a stream of crude mixture (which has the same ratio of A and B as the mixture in the still) can be added to the distilling mixture to replenish the liquid, meaning that the system can be run continuously.

General Improvements

Both batch and continuous distillations can be improved by making use of a fractionating column on top of the distillation flask. The column improves separation by providing a larger surface area for the vapor and condensate to come into contact. This helps it remain at equilibrium for as long as possible. The column can even exist of small subsystems ('dishes') which all contain an enriched, boiling liquid mixture, all with their own vapor phase.

There are differences between laboratory-scale and industrial-scale fractionating columns, but the principles are the same. Examples of fractionating columns (in increasing efficacy) include:

- Air condenser.
- Vigreux column (usually laboratory scale only).
- Packed column (packed with glass beads, metal pieces, or other chemically inert material).
- Spinning band distillation system.

Laboratory Scale Distillation

Laboratory scale distillations are almost exclusively run as batch distillations. The device used in distillation, sometimes referred to as a still, consists at a minimum of a reboiler or pot in which the source material is heated, a condenser in which the heated vapor is cooled back to the liquid state, and a receiver in which the concentrated or purified liquid, called the distillate, is collected. Several laboratory scale techniques for distillation exist.

Simple Distillation

In simple distillation, all the hot vapors produced are immediately channeled into a condenser which cools and condenses the vapors. Thus, the distillate will not be pure - its composition will be identical to the composition of the vapors at the given temperature and pressure, and can be computed from Raoult's law.

As a result, simple distillation is usually used only to separate liquids whose boiling points differ greatly (rule of thumb is 25 °C), or to separate liquids from involatile solids or oils. For these cases, the vapor pressures of the components are usually sufficiently different that Raoult's law may be neglected due to the insignificant contribution of the less volatile component. In this case, the distillate may be sufficiently pure for its intended purpose.

Fractional Distillation

For many cases, the boiling points of the components in the mixture will be sufficiently close that Raoult's law must be taken into consideration. Thus, fractional distillation must be used in order to separate the components well by repeated vaporization-condensation cycles within a packed fractionating column.

As the solution to be purified is heated, its vapors rise to the fractionating column. As it rises, it cools, condensing on the condenser walls and the surfaces of the packing material. Here, the condensate continues to be heated by the rising hot vapors; it vaporizes once more. However, the composition of the fresh vapors are determined once again by Raoult's law. Each vaporization-condensation cycle (called a theoretical plate) will yield a purer solution of the more volatile component. In reality, each cycle at a given temperature does not occur at exactly the same position in the fractionating column; theoretical plate is thus a concept rather than an accurate description.

More theoretical plates lead to better separations. A spinning band distillation system uses a spinning band of Teflon or metal to force the rising vapors into close contact with the descending condensate, increasing the number of theoretical plates.

Steam Distillation

Like vacuum distillation, steam distillation is a method for distilling compounds which are heat-sensitive. This process involves using bubbling steam through a heated mixture of the raw material. By Raoult's law, some of the target compound will vaporize (in accordance with its partial pressure). The vapor mixture is cooled and condensed, usually yielding a layer of oil and a layer of water.

Steam distillation of various aromatic herbs and flowers can result in two products; an essential oil as well as a watery herbal distillate. The essential oils are often used in perfumery and aromatherapy while the watery distillates have many applications in aromatherapy, food processing and skin care.

Vacuum Distillation

Some compounds have very high boiling points. To boil such compounds, it is often better to lower the pressure at which such compounds are boiled instead of increasing the temperature. Once the pressure is lowered to the vapor pressure of the compound (at the given temperature), boiling and the rest of the distillation process can commence.

This technique is referred to as vacuum distillation and it is commonly found in the laboratory in the form of the rotary evaporator.

This technique is also very useful for compounds which boil beyond their decomposition temperature at atmospheric pressure and which would therefore be decomposed by any attempt to boil them under atmospheric pressure.

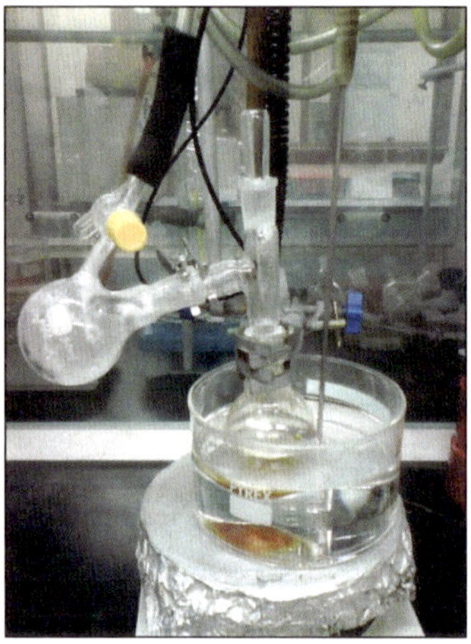

Dimethylsulfoxide usually boils at 189 °C. Under a vacuum, it distills off into the receiver at only 70 °C.

Air-sensitive Vacuum Distillation

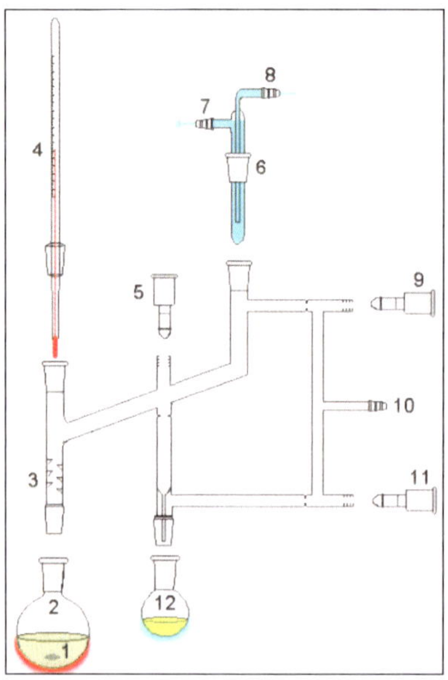

Perkin Triangle Distillation Setup.

In the figure, 1: Stirrer bar/anti-bumping granules 2: Still pot 3: Fractionating column 4: Thermometer/Boiling point temperature 5: Teflon tap 1 6: Cold finger 7: Cooling water out 8: Cooling water in 9: Teflon tap 2 10: Vacuum/gas inlet 11: Teflon tap 3 12: Still receiver.

Some compounds have high boiling points as well as being air sensitive. A simple vacuum distillation system as exemplified above can be used, whereby the vacuum is replaced with an inert gas after the distillation is complete. However, this is a less satisfactory system if one desires to collect fractions under a reduced pressure. To do this a "pig" adaptor can be added to the end of the condenser, or for better results or for very air sensitive compounds a Perkin triangle apparatus can be used.

The Perkin triangle has means via a series of glass or teflon taps to allow fractions to be isolated from the rest of the still, without the main body of the distillation being removed from either the vacuum or heat source, and thus can remain in a state of reflux. To do this, the sample is first isolated from the vacuum by means of the taps, the vacuum over the sample is then replaced with an inert gas (such as nitrogen or argon) and can then be stoppered and removed. A fresh collection vessel can then be added to the system, evacuated and linked back into the distillation system via the taps to collect a second fraction, and so on, until all fractions have been collected.

Azeotropic Distillation

Interactions between the components of the solution create properties unique to the solution, as most processes entail nonideal mixtures, where Raoult's law does not hold. Such interactions can result in a constant-boiling azeotrope which behaves as if it were a pure compound (i.e., boils at a single temperature instead of a range). At an azeotrope, the solution contains the given component in the same proportion as the vapor, so that evaporation does not change the purity, and distillation does not effect separation. For example, ethyl alcohol and water form an azeotrope of 95 percent at 78.2 °C.

If the azeotrope is not considered sufficiently pure for use, there exist some techniques to break the azeotrope to give a pure distillate. This set of techniques are known as azeotropic distillation. Some techniques achieve this by "jumping" over the azeotropic composition (by adding an additional component to create a new azeotrope, or by varying the pressure). Others work by chemically or physically remove or sequester the impurity. For example, to purify ethanol beyond 95 percent, a drying agent or a desiccant such as potassium carbonate can be added to convert the soluble water into insoluble water of crystallization. Molecular sieves are often used for this purpose as well.

Short Path Distillation

Short path distillation is a distillation technique that involves the distillate traveling a short distance, often only a few centimeters. A classic example would be a distillation involving the distillate traveling from one glass bulb to another, without the need for a condenser separating the two chambers. This technique is often used for compounds which are unstable at high temperatures. Advantages are that the temperature of the boiling liquid does not have to be much higher than the boiling point of the distilling substance, and the gases only have to travel a short distance while in the gas-phase before they can be cooled again to a lower temperature.

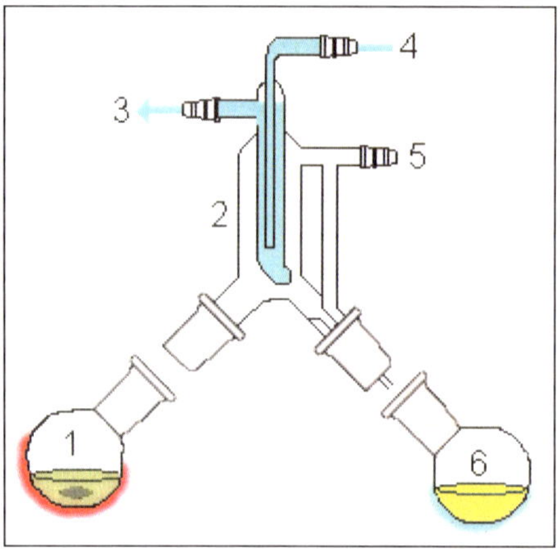

Short path vacuum distillation apparatus with vertical condenser (cold finger), to minimize the distillation path; 1: Still pot with stirrer bar/anti-bumping granules 2: Cold finger - bent to direct condensate 3: Cooling water out 4: cooling water in 5: Vacuum/gas inlet 6: Distillate flask/Distillate.

Other Types

- In rotary evaporation a vacuum distillation apparatus is used to remove bulk solvents from a sample. Typically the vacuum is generated by a water aspirator or a membrane pump.
- In a kugelrohr a short path distillation apparatus is typically used (generally in combination with a (high) vacuum) to distill high boiling (> 300 °C) compounds. The apparatus consists of an oven in which the compound to be distilled is placed, a receiving portion which is outside of the oven, and a means of rotating the sample. The vacuum is normally generated by using a high vacuum pump.
- The process of reactive distillation involves using the reaction vessel as the still. In this process, the product is usually significantly lower-boiling than its reactants. As the product is formed from the reactants, it is vaporized and removed from the reaction mixture. This technique is an example of a continuous vs. a batch process; advantages include less downtime to charge the reaction vessel with starting material, and less workup.
- Destructive distillation involves the strong heating of solids (often organic material) in the absence of oxygen (to prevent combustion) to evaporate various high-boiling liquids, as well as thermolysis products. The gases evolved are cooled and condensed as in normal distillation. The destructive distillation of wood to give methanol is the root of its common name - wood alcohol.

- Pervaporation is a method for the separation of mixtures of liquids by partial vaporization through a non-porous membrane.
- Dry distillation, despite its name, is not truly distillation, but rather a chemical reaction known as pyrolysis in which solid substances are heated in a strongly reducing atmosphere and any volatile fractions are collected.
- Extractive distillation is defined as distillation in the presence of a miscible, high boiling, relatively non-volatile component, the solvent, that forms no azeotrope with the other components in the mixture.
- Flash evaporation (or partial evaporation) is the partial vaporization that occurs when a saturated liquid stream undergoes a reduction in pressure by passing through a throttling valve or other throttling device. This process is one of the simplest unit operations.
- Freeze distillation is an analogous method of purification using freezing instead of evaporation. It is not truly distillation, and does not produce products equivalent to distillation. This process is used in the production of ice beer and ice wine to increase ethanol and sugar content, respectively.
- Codistillation is distillation which is performed on mixtures in which the two compounds are not miscible.

Industrial Distillation

Large scale industrial distillation applications include both batch and continuous fractional, vacuum, azeotropic, extractive, and steam distillation. The most widely used industrial applications of continuous, steady-state fractional distillation are in petroleum refineries, petrochemical and chemical plants and natural gas processing plants.

Typical industrial distillation towers.

Industrial distillation is typically performed in large, vertical cylindrical columns known as distillation towers or distillation columns with diameters ranging from about 65 centimeters to 6 meters and heights ranging from about 6 meters to 60 meters or more. When the process feed has a diverse composition, as in distilling crude oil, liquid outlets at intervals up the column allow for the withdrawal of different fractions or products having different boiling points or boiling ranges. The "lightest" products (those with the lowest boiling point) exit from the top of the columns and the "heaviest" products (those with the highest boiling point) exit from the bottom of the column and are often called the bottoms.

Large-scale industrial towers use reflux to achieve a more complete separation of products. Reflux refers to the portion of the condensed overhead liquid product from a distillation or fractionation tower that is returned to the upper part of the tower as shown in the schematic diagram of a typical, large-scale industrial distillation tower. Inside the tower, the downflowing reflux liquid provides cooling and condensation of the upflowing vapors thereby increasing the efficacy of the distillation tower. The more reflux is provided for a given number of theoretical plates, the better is the tower's separation of lower boiling materials from higher boiling materials. Alternatively, the more reflux is provided for a given desired separation, the fewer theoretical plates are required.

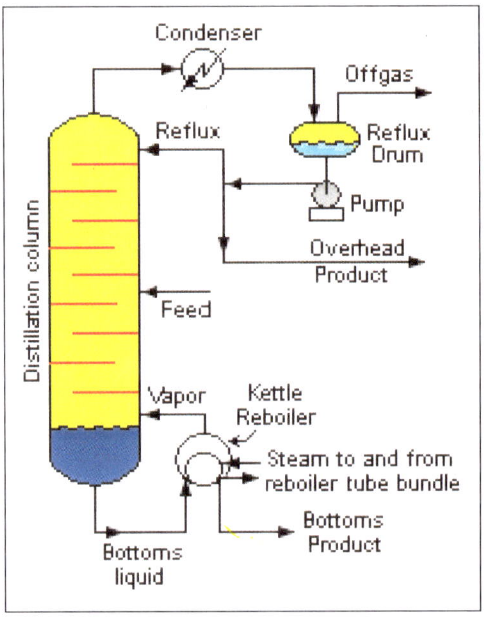

Diagram of a typical industrial distillation tower.

Such industrial fractionating towers are also used in air separation, producing liquid oxygen, liquid nitrogen, and high purity argon. Distillation of chlorosilanes also enables the production of high-purity silicon for use as a semiconductor.

Design and operation of a distillation tower depends on the feed and desired products. Given a simple, binary component feed, analytical methods such as the McCabe-Thiele method or the Fenske equation can be used. For a multi-component feed, simulation models are used both for design and operation. Moreover, the efficiencies of the vapor-liquid contact devices (referred to as "plates" or "trays") used in distillation towers are typically lower than that of a theoretical 100 percent efficient equilibrium stage. Hence, a distillation tower needs more trays than the number of theoretical vapor-liquid equilibrium stages.

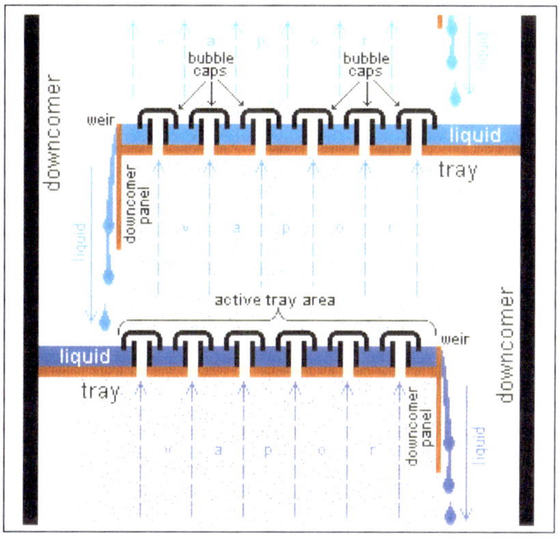

Section of an industrial distillation tower showing detail of trays with bubble caps.

In industrial uses, sometimes a packing material is used in the column instead of trays, especially when low pressure drops across the column are required, as when operating under vacuum. This packing material can either be random dumped packing (1-3" wide) such as Raschig rings or structured sheet metal. Liquids tend to wet the surface of the packing and the vapors pass across this wetted surface, where mass transfer takes place. Unlike conventional tray distillation in which every tray represents a separate point of vapor-liquid equilibrium, the vapor-liquid equilibrium curve in a packed column is continuous. However, when modeling packed columns, it is useful to compute a number of "theoretical stages" to denote the separation efficiency of the packed column with respect to more traditional trays. Differently shaped packings have different surface areas and void space between packings. Both of these factors affect packing performance.

Distillation in Food Processing

Distilled Beverages

Carbohydrate-containing plant materials are allowed to ferment, producing a dilute solution of ethanol in the process. Spirits such as whiskey and rum are prepared by distilling these dilute solutions of ethanol. Other components than ethanol are collected in the condensate, including water, esters, and other alcohols which account for the flavor of the beverage.

References

- Plawsky, Joel L. (April 2001). Transport phenomena fundamentals (Chemical Industries Series). CRC Press. Pp. 1, 2, 3. ISBN 978-0-8247-0500-8
- Chemical-reactors, processes: essentialchemicalindustry.org, Retrieved 21 July, 2019
- Bird, R.B., Stewart, W.E. and Lightfoot, E.N. (August 2001). Transport Phenomena (Second ed.). John Wiley & Sons. ISBN 0-471-41077-2
- Delgass; et al. «Seventy Five Years of Chemical Engineering». Purdue University. Retrieved 13 August 2013
- Truskey, George; Yuan F; Katz D. Transport Phenomena in Biological Systems (Second ed.). Prentice Hall. P. 888. ISBN 978-0131569881
- Arthur D. Little, William H. Walker, and Warren K. Lewis». Science History Institute. Retrieved 20 March 2018
- Distillation, entry: newworldencyclopedia.org, Retrieved 27 May, 2019
- Sinnott and Towler (2009). Chemical Engineering Design: Principles, Practice and Economics of Plant and Process Design (5th ed.). Butterworth-Heinemann. ISBN 0750685514

PERMISSIONS

All chapters in this book are published with permission under the Creative Commons Attribution Share Alike License or equivalent. Every chapter published in this book has been scrutinized by our experts. Their significance has been extensively debated. The topics covered herein carry significant information for a comprehensive understanding. They may even be implemented as practical applications or may be referred to as a beginning point for further studies.

We would like to thank the editorial team for lending their expertise to make the book truly unique. They have played a crucial role in the development of this book. Without their invaluable contributions this book wouldn't have been possible. They have made vital efforts to compile up to date information on the varied aspects of this subject to make this book a valuable addition to the collection of many professionals and students.

This book was conceptualized with the vision of imparting up-to-date and integrated information in this field. To ensure the same, a matchless editorial board was set up. Every individual on the board went through rigorous rounds of assessment to prove their worth. After which they invested a large part of their time researching and compiling the most relevant data for our readers.

The editorial board has been involved in producing this book since its inception. They have spent rigorous hours researching and exploring the diverse topics which have resulted in the successful publishing of this book. They have passed on their knowledge of decades through this book. To expedite this challenging task, the publisher supported the team at every step. A small team of assistant editors was also appointed to further simplify the editing procedure and attain best results for the readers.

Apart from the editorial board, the designing team has also invested a significant amount of their time in understanding the subject and creating the most relevant covers. They scrutinized every image to scout for the most suitable representation of the subject and create an appropriate cover for the book.

The publishing team has been an ardent support to the editorial, designing and production team. Their endless efforts to recruit the best for this project, has resulted in the accomplishment of this book. They are a veteran in the field of academics and their pool of knowledge is as vast as their experience in printing. Their expertise and guidance has proved useful at every step. Their uncompromising quality standards have made this book an exceptional effort. Their encouragement from time to time has been an inspiration for everyone.

The publisher and the editorial board hope that this book will prove to be a valuable piece of knowledge for students, practitioners and scholars across the globe.

INDEX

M

N

O

P

R

S

T

U

V